Chemistry of Volumetric - Water Quality Analysis

Chemistry of Volumetric - Water Quality Analysis

P.V. NIDHEESH
Senior Scientist
CSIR - National Environmental
Engineering Research Institute (NEERI)
Nagpur - 440 020 (Maharashtra)

SATISH SERIAL PUBLISHING HOUSE
403, Express Tower, Commercial Complex
Azadpur, Delhi-110033 (India)
Phone : 011-27672852, Fax : 91-11-27672046
e-mail : info@satishserial.com, hkjain1975@yahoo.com
Website : www.satishserial.com

Published by :

SATISH SERIAL PUBLISHING HOUSE

403, Express Tower, Commercial Complex, Azadpur, Delhi-110033 (INDIA)
Phone : 011-27672852 Fax : 91-11-27672046
E-mail : info@satishserial.com, hkjain1975@yahoo.com

ISBN 978-93-86200-08-2

Composed, Designed & Printed in India

Preface

Due to our increased day to day needs and other activities, the deterioration of surface and ground water quality occurs rapidly. The latest study by WHO and UNICEF reports that, more than 663 million people in the world are using contaminated water for drinking purpose. Most of the water contaminants can be quantified by volumetric analysis.

Since, water quality is an important factor; most universities included its quantification methods in their curriculum. The chemistry behind this volumetric water quality analysis is very important for determining the contaminant concentration correctly. Most teachers, students and consultants lack the exact water contaminant quantification due to the unavailability of a standard book.

An Environmental Engineering lab generally contains several acids, bases, salts etc. Safety measures undertaken while handling these chemicals is of prime importance . This book also gives a general idea on the safety measures to be taken before the usage of various chemicals in the lab.

Author

Acknowledgement

First of all, I feel grateful to thank Director, CSIR-NEERI, Nagpur for his encouragement and guidance to publish this book. I wish to express my profound thanks to all participants, Indian Society of Technical Education (ISTE) approved training programme on "Water Pollution: Contaminant Quantification and Treatment" held at Vimal Jyothi Engineering College, Chemperi, Kannur Kerala during 30th November 2015 to 4th December 2015; for motivating me to write this book. With a deep sense of gratitude, I wish to thanks to my teachers, friends, co-workers and family members. I will be grateful to the readers if they notify the portions where improvements are required.

Dr. P.V. Nidheesh

Contents

1

Safety in Environmental Engineering Lab

Since, various types of chemicals are used in an Environmental Engineering laboratory, safety of persons working in the lab is of great concern. For the efficient functioning of lab and the safety of workers, the following points should be considered [1].

- Before and after the usage, clean the glassware thoroughly.
- Handle the glassware carefully.
- Always add acid to water, not water to acid.
- Use different pipette for different reagents.
- Don't pipette out acids and other toxic reagents by mouth.
- Use large volumes of water to prevent serious acid burns.
- Don't handle chemicals with bare hands.

Apart from the above points, handling of chemicals demands strict attention. Following data lists the precautions to be taken while handling chemicals. The list summarises only the common chemicals in the laboratory.

Hydrochloric acid, Sulphuric acid, Nitric acid [2-4]**:** These acids are very common in an environmental engineering lab. These acids are very toxic and hazardous in nature. The carcinogenic effect of sulphuric acid and hydrochloric acid has been proven. All the above acids are corrosive in nature. These acids are toxic to kidneys, liver, mucous membranes, upper respiratory tract, skin, eyes, circulatory system and teeth. These acids are very hazardous in case of skin contact (corrosive, irritant, permeate lining), eye contact (irritant, corrosive), ingestion; and slightly hazardous in case of inhalation (lung sensitizer). Repeated exposure to these acids may cause target organs damage, chronic eye irritation, severe skin irritation, respiratory tract irritation, general deterioration of health etc. Breathing of the fumes produced from these acids is also very dangerous.

Apart from toxic, carcinogenic and hazardous nature, these acids are hygroscopic. Therefore, special care should be taken while handling the acids. These acids must be kept in a dry and tightly closed container in a cool, well-ventilated area. The acids must be stored below 23°C. The store should be kept away from oxidizing agents, reducing agents, combustible materials, organic materials, metals, acids, alkalis, moisture etc.; and should be locked.

Sodium Hydroxide [5]**:** Sodium hydroxide is the commonly used base. This chemical may be toxic to mucous membranes, upper respiratory tract, skin, eyes. But, sodium hydroxide is very hazardous in case of skin contact (corrosive, irritant, permeate skin), eye contact (irritant, corrosive), ingestion, inhalation. Corneal damage or blindness may occur by the eye contact. The contact of skin with this alkali can cause inflammation and blistering. Repeated skin contact may lead to local skin destruction, or dermatitis. Similarly, the repeated inhalation of dust can produce respiratory irritation or lung damage.

Similar to sulphuric acid, hydrochloric acid and nitric acid; sodium hydroxide is also hygroscopic in nature. Therefore, the substance should be kept in a dry container and in a cool, well- ventilated area. Don't add water to sodium hydroxide. This alkali must be kept away from oxidizing agents, reducing agents, metals, acids, alkalis and moisture.

Silver nitrate and potassium chromate [6, 7]**:** These chemicals are the main reagents in chloride determination test by Mohr's method. Both chemicals (silver nitrate, potassium chromate) have chronic health effects. Silver nitrate is toxic to lungs and may also affect mucous membranes, skin and eyes. Toxicity of potassium chromate is less than that of silver nitrate. Potassium chromate may be toxic to kidneys, lungs, liver, upper respiratory tract, skin, eyes. Both chemicals are very hazardous in case of eye and skin contacts and ingestion.

Since, potassium chromate and silver nitrate are hazardous in nature; a few precautions should be taken for handling these chemicals. Keep these chemicals away from heat, combustible material and sources of ignition. Keep the tightly closed containers of these chemicals in a cool, well-ventilated area. Store these chemicals separately, especially from acids, alkalies, reducing agents and combustibles.

Acetic acid [8]**:** Acetic acid is hazardous and toxic in nature. This acid is very hazardous in case of skin contact (irritant), eye contact (irritant), ingestion and inhalation. Toxicity of the acid is significant to kidneys, mucous membranes, skin, and teeth. Organs damage may happen due to the repeated or prolonged exposure to acetic acid. Due to the hazardous and toxic nature of the acid, the acid must be kept in a secluded and approved area. The tightly closed containers of the acid must be stored in a cool and ventilated place. The container should be away from all possible sources of ignition. Don't breathe the fumes.

Bleaching powder [9]**:** Bleaching powder is an unstable material, which may cause fire on contact with combustible materials. This substance may cause skin burning and eye irritation. Inhalation of the chlorine gas produced from bleaching powder irritates nose, throat and upper respiratory tract. Therefore, this substance should be kept away from heat, sparks and flames in a well-ventilated area. In case of contact with skin, wash immediately with plenty of water.

FIRST AID MEASURES

Eye contact of chemicals cause serious problems. In case of eye contact, flush the eyes with plenty of water for at least 15 min. Cold water is better to use for flushing. Repeat flushing eyes, if the irritation persists even after 15 min flushing.

In the case of skin contact, wash the skin with cold water for at least 15 min. Remove the contaminated clothing and shoes. Wash affected area with a disinfectant soap and cover with an anti-bacterial cream. Get medical attention immediately for the contact of acids, alkali, bleaching powder, etc.

Inhalation of fumes produced from acids, gas vapours produced from bleaching powder and other chemicals cause serious health problems. Inhalation of these chemicals reduces the oxygen content in our blood. Loosen tight clothing such as a collar, tie, belt or waistband. Move the victim to fresh air and give artificial respiration (mouth-to-mouth resuscitation) if not breathing properly. If breathing is difficult, provide oxygen to the victim. Seek immediate medical attention.

In the case of ingestion, rinse mouth of victim with cold water thoroughly. Give plenty of water to the victim and this will dilute the swallowed material. Seek immediate medical attention. Don't induce vomiting unless directed to do so by medical personnel. If spontaneous vomiting occurs, make victim lean forward with head down to avoid breathing in of vomit, rinse mouth and administer more water. Never give anything by mouth to an unconscious person.

FURTHER READING

Material safety data sheets of other chemicals.

❑❑❑

2

Fundamental Chemistry of Volumetric Analysis

VOLUMETRIC ANALYSIS

Volumetric or titrimetric analysis is a quantitative method which employ normality equation for finding the concentration of a solution of known volume by titrating with a standard solution of known volume and concentration. Suppose the concentration of a solution "X" is to be finding out after the titration with a standard solution "Y". The standard solution "Y" added into the solution "X" incrementally until the concentration of "X" has been consumed. Volume of standard solution consumed for the reaction is measured and is equivalent to the concentration of "X". This is known as equivalence point and it is very difficult to identify the equivalence point in the absence of an external source. In most of the volumetric analysis, a chemical compound which is capable of indicating the equivalence point is added. This compound is known as indicators and it undergoes rapid change of physical property, usually colour, at equivalence point. Some of the indicators react with the titrant to indicate the equivalence point. For example, potassium chromate indicator used for chloride determination test reacts with silver nitrate (which is the titrant) and form a pinkish yellow precipitate of silver chromate. Some of the indicator gives equivalence point without reacting with the titrant. This type of indicator gives colour change according to the characteristics of solution. For example, phenolphthalein indicator shows pink colour at alkaline condition and is colourless at pH around 8.3.

The following components are essential to complete the volumetric analysis successfully.

1. The reaction equation must be known. This helps to find the concentration of unknown solution stoichiometrically.

2. A rapid completion of the reaction is a must. The volumetric analysis is useful only when the reaction is completed within a short period.
3. A pre-treatment to remove interferents.
4. An indicator which shows the end point of the reaction accurately.
5. A device to measure the volume of titrant (burette) and water sample (measuring cylinder, pipette, volumetric flak, analytical balance).

FUNDAMENTAL CHEMISTRY

The volumetric analysis of water is carried out for solutions. A solution is a mixture of two or more substances, in which one or more components are homogeneously dissolved in another component. In a solution, the quantity of one component which is always higher than all other components, is known as solvent. All other components are dissolved in the solvent. The other component(s) of the solution other than the solvent is called solute. Thus a solution is formed by dissolving the solute uniformly in a solvent. Consider a sugar solution as an example. In this, the sugar is the solute and water is the solvent.

The physical state of the solvent determines the physical state of the solution. There are three types of solutions: gaseous, liquid and solid solutions. In all cases, the physical state of the solvent and the solution is same. But the physical state of solute may vary. For example, liquid solutions can be prepared by dissolving gas, liquid or solid in a liquid solvent. Thus a liquid solution can be prepared by dissolving oxygen in water, ethanol in water or sodium chloride in water.

The solubility of solute in the solvent is an important parameter which determines the homogeneity of the solution. The solubility of a solute is defined as the amount of solute dissolved in the solvent. The solubility is dependent on the nature of solute, nature of solvent, temperature, etc. Some of the solute is easily soluble and others are not. For example, sodium chloride dissolves rapidly in water. At the same time, most of the gases like oxygen, chlorine etc. are sparingly soluble in water.

There is a limit for the quantity of solute which can be dissolved in a solvent and this upper solute concentration is known as saturation concentration. The dissolved form and the undissolved form of the solute are always in equilibrium in a saturated solution. Therefore, two processes occur in a saturated solution: dissolution and crystallisation. Further increase in solute concentration enhances the rate of crystallisation.

CONCENTRATION OF A SOLUTION

The concentration of a solution can be expressed in many ways. The following are some of the expressions [10].

Mass percentage (w/w): The concentration of solute is expressing as a percentage of total weight of solution. It is expressed as:

$$\text{Mass percentage (\%)} = \frac{\text{Weight of solute (g)}}{\text{Weight of solution (g)}} \times 100 \quad (2.1)$$

For example, 5% NaCl means every 100 g of solution contains 5 g of NaCl.

Volume percentage (v/v): The concentration of solute is expressed as a percentage of total volume of solution. It is expressed as:

$$\text{Volume percentage (\%)} = \frac{\text{Volume of solute (mL)}}{\text{Volume of solution (mL)}} \times 100 \quad (2.2)$$

For example, 5% of ethyl alcohol solution in water means that 5 mL of ethyl alcohol is dissolved in 95 mL of water.

Parts per million (ppm): It is defined as:

$$\text{ppm} = \frac{\text{Number of parts of the component}}{\text{Total number of parts of all components in the solution}} \times 10^6 \quad (2.3)$$

Generally, ppm is expressed when the solute is present in trace amount. One ppm is equal to one mg/L. Thus one ppm solution can be prepared by dissolving one gram of solute in one litre of solvent.

Molarity (M): It is the number of moles of solute dissolved in one litre of solution.

$$\text{Molarity (M)} = \frac{\text{Moles of solute}}{\text{Volume of solution}}$$

$$= \frac{\text{Weight of solute (g)}}{\text{Molecular weight of solute (g) x Volume of solution (L)}} \quad (2.4)$$

For example, the molarity of 200 mL solution containing 2 g of sodium chloride

$$= \frac{2\text{g}}{58.5 \text{ g} \times 0.2\text{L}} = 0.17 \text{ M}$$

Normality (N): It is the number of equivalent weights of solute dissolved in one litre of solution.

$$\text{Normality (N)} = \frac{\text{Weight of solute (g)}}{\text{Equivalent weight of solute (g) x Volume of solution (L)}} \quad (2.5)$$

Equivalent weight is the number of grams of a substance that will provide Avogadro's number of units of reaction. The equivalent weight is

$$\text{Equivalent weight} = \frac{\text{Molecular weight (g)}}{C_c} \quad (2.6)$$

Where C_c is the combining capacity of the solute. The value of C_c for acids, bases and salts is equal to the number of protons, number of hydroxyl ions and total number of positive or negative charges, respectively. In the case of redox reactions, the value of C_c is equal to the number of electrons transferred in the reaction or change in oxidation number. This equation is mainly applicable for heavy metals.

Q1. Find the equivalent weight of (a) $CaCO_3$ (b) H_2SO_4 (c) $Ca(OH)_2$ (d) $FeSO_4$ (e) $K_2Cr_2O_7$

Ans. a) Calcium carbonate is a salt. It dissociates as

$$CaCO_3 \longrightarrow Ca^{2+} + CO_3^{2-}$$

C_c= Total number of positive charges = total number of negative charges = 2

Molecular weight of calcium carbonate = 100g

Equivalent weight = 100g/2 = 50g.

b) It is an acid. Therefore, C_c = number of protons = 2

Molecular weight = 98 g

Equivalent weight = 98/2 = 49g.

c) It is a base. Therefore, C_c = number of hydroxyl ions = 2

Molecular weight = 74 g

Equivalent weight = 74/2 = 37g.

d) In ferrous sulphate, iron is in its +2 states. Since, it is a heavy metal; the combining capacity is equal to the number of electrons transferred or change in the oxidation state. Iron has two oxidation states +2 and +3. Ferrous ion changes to ferric ion by the redox reaction. Therefore, the change in oxidation state or electrons transferred in this reaction is equal to one. Thus, the equivalent weight of ferrous sulphate is equal to its molecular weight (152g).

e) After the chemical reaction, one mole chromium in 6+ form in the potassium dichromate converts to two moles of its 3+ forms. Thus change in oxidation number or change in electrons is 6. Thus, equivalent weight of potassium dichromate is 1/6 of its molecular weight (i.e. 49.03 g).

Q2. What is the normality of 200 mL of a solution containing 2 g of $CaCO_3$?

Ans. Weight of solute = 2g; Equivalent weight of solute = 50 g; Volume of solution = 0.2 L

Normality = 2/(50 x 0.2) = **0.2 N**

DILUTION EQUATIONS

The solution can be diluted by adding solvent into the solution. Thus the amount of solute in the diluted solution is same. But its concentration changes after the dilution. Following relationships are valid for the dilution.

$$M_1 V_1 = M_2 V_2 \qquad (2.7)$$

$$C_1 V_1 = C_2 V_2 \qquad (2.8)$$

$$N_1 V_1 = N_2 V_2 \qquad (2.9)$$

Where, M_1, N_1 and C_1 respectively are the molarity, normality and concentration (mg/L) of concentrated solution having the volume V_1 (L). M_2, N_2 and C_2 respectively are the molarity, normality and concentration (mg/L) of diluted solution having the volume V_2 (L).

Q3. Find the volume of concentrated sulphuric acid (M.W. =98; 1 L = 1.84 kg) required to make (a) 500 mL of 0.2 N solution (b) 250 mL of 1 M solution and (c) 750 mL of 50 mg/L solution

Ans. Consider 1L of concentrated acid. Its weight is 1.84 kg

Normality of concentrated acid (N_1) = 1.84 x10^3/(49x1)=37.5 N

Molarity of concentrated acid (M_1) = 1.84 x 10^3/(98x1)=18.75 M

Concentration of concentrated acid (C_1) = 1840 g/L

a) Required normality (N_2) = 0.2 N

Required volume (V_2) = 0.5 L

$N_1V_1 = N_2\ V_2$

$37.5\ V_1 = 0.2 \times 500$

$V_1 = 2.67$ mL

Thus, 0.2 N sulphuric acid can be prepared by diluting 2.67 mL of concentrated acid and make up to 500 mL.

b) Required molarity (M_2) = 1 M

Required volume (V_2) = 0.25 L

$M_1V_1 = M_2\ V_2$

$18.75\ V_1 = 1 \times 250$ mL

$V_1 = 13.33$ mL

Thus, 1 M sulphuric acid can be prepared by diluting 13.33 mL of concentrated acid and make up to 250 mL.

c) Required concentration (C_2) = 50 mg/L

Required volume (V_2) = 0.75 L

$C_1V_1 = C_2\ V_2$

$1840 g/L \times V_1 = 50\ mg/L \times 750$ mL

$V_1 = 0.02$ mL

Practically it is very difficult to prepare 750 mL of 50 mg/L solution directly from the concentrated acid in a single step. Because, taking 0.02 mL is very difficult. This problem can be solved by preparing the solution in multi-steps. For example, prepare the 500 mL of solution having concentration 5g/L by diluting 1.35 mL of concentrated acid. Thus, 750 mL of 50 mg/L acid solution can be prepared by diluting 7.5 mL of 5g/L acid solution.

Q4. Find the volume of HCl (Purity 35%; 1L= 1.16 kg) required to make (a) 500 mL of 0.2 N solution (b) 250 mL of 1 M solution (c) 750 mL of 500 mg/L solution

Ans. Generally, 35% shows the weight percentage of HCl in water medium.

a. Consider 1L of the acid solution

The amount of HCl in this solution = 1.16 x 0.35 = 0.406 kg = 406g

Therefore, normality of this solution = 406/36.5 = 11.12 N

We have, $N_1V_1 = N_2 V_2$

$11.12 V_1 = 0.2 \times 500$

$V_1 = 9$ mL

Thus, 0.2 N HCl solution can be prepared by diluting 9 mL of 35% HCl solution to 500 mL.

b. Since, the number of protons is equal to one; normality and molarity of the solution are same.

$M_1V_1 = M_2 V_2$

$11.12 V_1 = 1 \times 250$

$V_1 = 22.48$ mL

Thus, 1M HCl solution can be prepared by diluting 22.48 mL of 35% HCl solution to 250 mL.

c. $C_1V_1 = C_2 V_2$

406 g/L V_1 = 500 mg/L x 750 mL

$V_1 = 0.93$ mL

Thus, 500 mg/L HCl solution can be prepared by diluting 0.93 mL of 35% HCl solution to 750 mL. Since the required volume is very less, this can be prepared in multi-steps.

❑❑❑

3

Chloride

INTRODUCTION

Chlorides are present in natural water mainly in the form of sodium chloride. River water and groundwater contain considerable amount of chloride. At the same time, the chloride content present in upland and mountain supplies is very less compared to other sources. This indicates that the chloride content in water depends on the mineral composition present in that area. During the run-off the minerals present in the earth surface dissolve in the water. This results in the increased concentration of salt in river water. The salts present in soil and rocks dissolve in infiltrated water during percolation. This increases the salt concentration of ground water. The concentration of chloride in water increases with increase in mineral concentration.

The chloride content in the sea and ocean water is very high compared to other surface and ground water. Around 3.5% of sea water contains salt. This increased salt concentration in sea and ocean water is the result of evaporation of natural water flowing into the sea or ocean. The salt remains in sea as a residue of the evaporation process. Sometimes, the salt concentration of lakes is higher than that of sea. For example, the salt concentration in the Great Salt Lake, USA has a salt concentration in the range of 23% [11]. Lakes are shallow than sea and this increases the higher evaporation rate in lake compared to sea. Thus, lakes, especially land-locked lakes without known outlets are more salty than an ocean.

Chloride content in the natural water also increases with increase in water pollution from other sources. The chloride content in water increases with increase in sea water intrusion and leaching from marine sedimentary deposits, solid waste dumping site and industrial wastes etc. Thus, higher chloride concentration in water indicates higher water pollution.

Higher chloride content in drinking water will not cause any problem to human health. Water having higher chloride content even in the range of 2 g/L is used for domestic purpose in water scared area. Thus, the human health is not adversly affected once adapted with higher chloride content. But the cation associated with the chloride has various adverse effects. Sodium chloride imparts salty taste to the water, if the concentration of chloride is higher than 250 mg/L. Thus, the desirable chloride content in drinking water is limited to 250 mg/L [12]. Calcium chloride and magnesium chloride cause permanent hardness to water. Heating of magnesium chloride generates hydrochloric acid. This is also highly corrosive and creates problems in boilers.

Chloride is normally used as a tracer for various environmental studies. A tracer is any matter or energy carried by water which will give information concerning the direction and or velocity of water as well as potential contaminants which could be transported by the water [13]. Chloride is a very efficient hydrological tracer. This is mainly due to the following reasons [14].

- Chloride ions can be easily detected and measured from water medium
- Chloride ions are not adsorbed by soils easily.
- It is a non-biodegradable ion
- Its presence is not visually detectable
- Non-toxic in nature

Thus, leaching from a septic tank to groundwater can be identified by chloride determination. Microbial analysis is the most suitable method for the determination of groundwater pollution by the leaching from a septic tank. But, these studies are time consuming. This can be overcome by chloride analysis. Because, human excreta especially urine contain high amounts of chloride. The chloride content in urine is equal to the chloride content in the water and food consumed by the human being. Thus the chloride content in the urine is in the range of 6g per person per day and this increases the municipal wastewater chloride content up to 15 mg/L [14]. Thus the leaching from the septic tank also increases the chloride content of natural receiving water considerably.

Chloride attack on concrete is a severe problem in marine areas. Since chloride concentration is very high in marine regions, the chloride ions diffuse into concrete and extract calcium in the form of calcide. This reduces the strength of concrete.

The chloride content in irrigation water also needs to be controlled. The higher salt concentration in the water increases the salt concentration near the root zone due to the evapotranspiration. This higher salt concentration makes crops to take the water due to the osmotic pressure differences between the water outside the plants and within the plant cells. Thus it is advisable to provide water having lesser salt concentration, even below the drinking water standards for irrigating crops.

Rate of pumping groundwater in areas where sea water intrusion is suspected, also needs to be controlled with the chloride concentration. The concentration of chloride increases with increase in pumping rate or extension of pumping. Continuous chloride determination is required in such areas while pumping groundwater.

The type of desalting apparatus or process selection is also dependent on the chloride or total dissolved solids (TDS) concentration. Ion exchange, electrodialysis, nanofiltration and reverse osmosis is applicable for the salty water having TDS concentration up to 500 mg/L, 3000 mg/L, 10000 mg/L and above 10000 mg/L, respectively [11]. Distillation process is very suitable for the salty water having TDS concentration greater than 30000 mg/L [11].

DETERMINATION OF CHLORIDE BY MOHR'S METHOD

Principle

Mohr method is a widely used volumetric analysis method for the determination of chloride ion concentration in water. This method is also known as Argentometric method. The basic principle behind this analysis is the precipitation of silver chloride with the addition of silver ions to water containining chloride ions. Silver ions react quickly with halide ions and forms silver halide precipitate. Silver nitrate is used as the titrant for this analysis. With the addition of silver nitrate, the chloride ions present in the water reacts with silver ions and forms silver chloride precipitate as in Eq. 3.1. But, the silver chloride precipitate is a white coloured precipitate and hence it is difficult to identify the end point. Therefore, an indicator capable of indicating the presence of excess silver ions after the precipitation of silver chloride is required. Potassium chromate is normally used as the indicator in

Mohr method. The excess silver ions react with chromate ions and produce a reddish brown precipitate of silver chromate as in Eq. 3.2. Thus, potassium chromate indicator gives an end point of reddish brown from yellow. The yellow colour with the addition of potassium chromate is the colour of indicator.

$$Ag^+ + Cl^- \rightarrow AgCl \tag{3.1}$$

$$2Ag^+ + CrO_4^{2-} \rightarrow Ag_2CrO_4 \tag{3.2}$$

Reagent Preparation

Standard silver nitrate solution of 0.0141 N or 0.0282 N is used as titrant for the analysis. Following gives the procedure for the preparation of 0.0141 N silver nitrate solution. First, dissolve 2.395 g of silver nitrate in 500 mL distilled water and make up to 1L in a volumetric flask. This solution should be stored in an amber coloured bottle because the decomposition of silver nitrate occurs in the presence of sun light. Standardization of this solution can be carried out using the 0.0141 N NaCl solution.

Potassium chromate indicator can be prepared by dissolving 50g of potassium chromate in a little amount of distilled water, say 100 mL. Then, add silver nitrate solution drop by drop until red precipitate of silver chromate formed. Keep the solution for 12 hr. Filter the solution and dilute to 1 L with distilled water.

Standard Procedure and Chemistry behind the Analysis

1. Take V mL of the given sample in a conical flask.
2. Add 3 mL of aluminium hydroxide to this solution, if the sample is coloured. Then shake well, filter, wash and collect filtrate.

Aluminium hydroxide is a well known coagulant. The addition of aluminium hydroxide removes the impurities which impart colour to the water by the coagulation mechanism known as sweep floc coagulation or enmeshment in precipitate.

3. Add 1 mL H_2O_2 to the solution, if sulphide, sulphite, or thiosulphate is present in the sample. Then stir the solution for 1 min.

Sulphide, sulphite and thiosulphate are the major interferences for this analysis. These anions react with silver ions and form insoluble compounds. For example, silver reacts with thiosulphate and form silver thiosulphate as in the Eq. 3.3 [15]. Even though, the resultant of the above reaction is an insoluble compound, it then reacts with water at a higher rate and forms sulphuric acid and silver sulphide as in Eq. 3.4 [15]. These side reactions consume more amount of silver nitrate than the required amount during the analysis. Thus the presence of the above anions gives high chloride concentration after the volumetric analysis. Therefore, the removal of these anions before the analysis is necessary.

$$S_2O_3^{2-} + 2Ag^+ \rightarrow Ag_2S_2O_3 \quad (3.3)$$

$$Ag_2S_2O_3 + H_2O \rightarrow AgS + H_2SO_4 \quad (3.4)$$

These anions can be removed from the water by the addition of adequate amount of hydrogen peroxide. For example, with the addition of hydrogen peroxide, the thiosulphate ions present in the water reacts with hydrogen peroxide and forms sulphuric acid as in Eq. 3.5 [16].

$$S_2O_3^{2-} + 4H_2O_2 \rightarrow H_2SO_4 + 3H_2O + SO_4^{2-} \quad (3.5)$$

4. Check the pH of the solution. Make the pH of the solution 7 to 8 by adding acid or alkali.

This is an important step in the Mohr's analysis. The pH of the solution should be in between 7 to 8. At highly alkaline conditions, silver ions precipitated as silver hydroxide. Therefore, the reaction between chloride ions will not take place in a proper manner at elevated pH conditions. This results in an additional consumption of silver nitrate than the required quantity. In contrary to this, at acidic conditions, the chromate ions convert into dichromate ions. Thus, the pinkish yellow precipitate of silver chromate will not form at lower pH conditions. Therefore, the analysis at acidic conditions will not show the end point even after the precipitation of silver chloride.

For adjusting the pH, use sulphuric acid and sodium hydroxide solutions. Don't use hydrochloric acid for adjusting the solution pH because hydrochloric acid addition will impart additional chloride ions to the water.

5. Add 1 mL of indicator (Potassium chromate) to get light yellow colour. The light yellow colour is the colour of potassium chromate. The amount of indicator added in the water sample

should be very less. Even though the standard procedure is given as 1 mL, one or two drops of indicator are enough. Because, the addition of higher amount of potassium chromate accelerates the formation of silver chromate (as per Le Chatelier's principle). Thus the addition of higher amounts of indicator may show the presence of silver chromate before the actual end point reaches.

6. Titrate the solution against standard silver nitrate solution until a reddish brown (pinkish yellow, brick red) precipitate is obtained.

The formation of pinkish yellow precipitate indicates the formation of insoluble silver chromate in water medium. Depending on the concentration of chloride ions present in the water, the quantity of white silver chloride precipitate will vary. If the concentration of chloride ions in the water is very high, the white coloured silver chloride precipitate formation in the water sample is visible during the analysis.

7. Note down the volume of silver nitrate added (V_1).
8. Repeat the procedure for blank and note down the volume (V_2).

Blank correction is an important step in this analysis. The rate of silver chloride formation is very much higher than that of silver chromate [14]. Thus, silver chloride precipitates before the silver chromate formation. Since, the rate of formation of silver chromate is too low, additional silver ions are required for the effective precipitation. In order to measure this additional amount of silver ions, blank correction is required. Thus, the blank correction should be carried out using 'chloride free distilled water'. The volume of 'chloride free distilled water' for the analysis is same as that of the 'water sample' volume.

9. Calculate the amount of chloride present using following equation

$$\text{Amount of chloride present in the given sample mg/L} = \frac{(V_1 - V_2) \times \text{Normality of } AgNO_3 \times 35.45 \times 1000}{\text{Volume of sample taken (mL)}} \quad (3.6)$$

The overall chloride analysis procedure is depicted in Fig. 3.1.

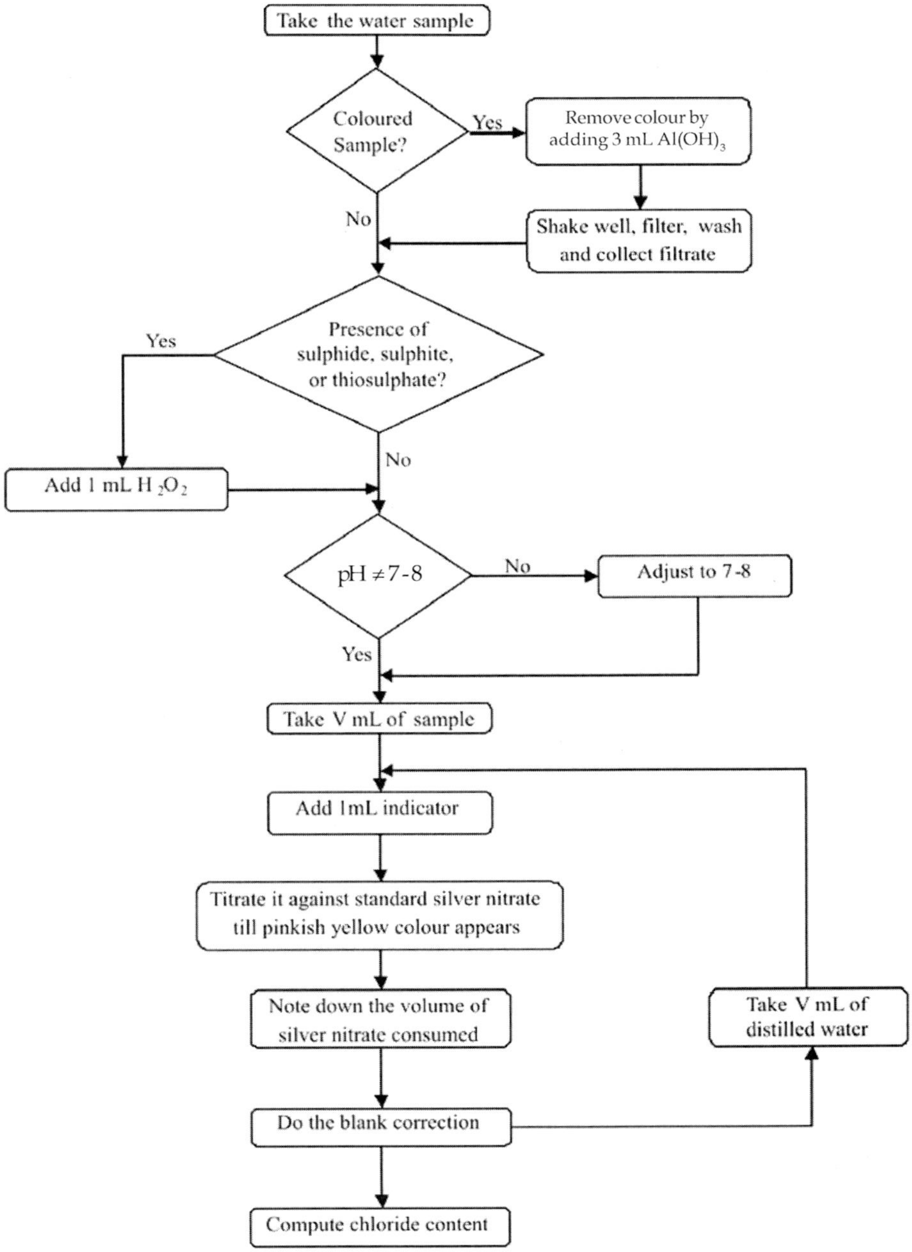

Fig. 3.1 Flowchart of chloride analysis procedure

TABULATION

Burette solution: Silver Nitrate (0.0141 N)

Indicator: Potassium Chromate

End Point: Yellow to reddish brown/ pinkish yellow

Titration 1 (Sample)

No.	Volume of Sample (mL)	Burette Reading (mL)		Volume of $AgNO_3$, V_1 (mL)
		Initial	Final	

Titration 2 (Chloride Free Distilled Water)

No.	Volume of Sample (mL)	Burette Reading (mL)		Volume of $AgNO_3$, V_2 (mL)
		Initial	Final	

QUESTIONS

1. What are the precautions to be taken during the analysis? Why?
2. Blank correction is necessary for Mohr's analysis. Why?
3. What is the pH of water required for the determination of chloride ions? Why?
4. During the volumetric analysis, the amount of potassium chromate added in the second titration is higher than that in the first titration. Comment on this.
5. The chloride concentration is sewage is always higher than that of water. Why?

6. Why is silver nitrate solution stored in amber coloured bottles?
7. Chloride is an efficient tracer. Why?
8. The chloride content in irrigation water should be very less. Why?
9. Explain the interferences caused by sulphide, sulphite and thiosulphite for the determination of chloride.

OBJECTIVE QUESTIONS

1. The permissible limit of chloride concentration for drinking water as per IS 10500:2012 is ....
2. The desirable limit of chloride concentration for drinking water as per IS 10500:2012 is ....
3. Another name of Mohr's method is......................................
4. The colour of silver chromate is..
5. The average salt concentration of sea water is....................................
6. The chloride concentration of sewage is .... of water.
7. Sometimes chloride content in .... is higher than that of sea.
8. Heating of .... generates hydrochloric acid.
9. Chloride is normally used as a .... for various environmental studies.
10. ....is main interferent in the determination of chemical oxygen demand.
11. Distillation process is suitable for salty water having TDS concentration greater than ....
12. Chloride extract calcium from concrete in the form of ....
13. The chloride content in the urine is in the range of ....
14. The interferences of sulphide, sulphite, or thiosulphate during chloride determination can be rectified by adding .... to it.
15. The ideal pH for the determination of chloride is ....

16. The rate of .... formation is very much higher than that of silver chromate
17. .... gives salty taste to water
18. Blank correction should be carried out using ....

4

Alkalinity

INTRODUCTION

The acid neutralizing capacity of water is termed as alkalinity. In other words, alkalinity is a measurement of ability of water to accept protons. The alkalinity of water is due to the presence of salts of weak acids and strong bases. These substances act as buffers and resist the drop of pH [14]. The alkalinity of water is mainly caused by three anions:- hydroxyl, carbonate and bicarbonate ions. Thus the alkalinity data is useful to measure the concentration of hydroxyl, carbonate and bicarbonate ion concentrations in water.

Higher and lower alkalinity in water will cause some severe problems. Higher amount of alkalinity imparts bitter taste to the water and is usually unpalatable. The water with higher alkalinity can dry out the skin and is responsible for scale formation. At the same time, the water with lower alkalinity is very corrosive and has no capacity to resist the changes in solution pH. The alkalinity of water can be measured using the Eq. 4.1. Generally, alkalinity is expressed as $CaCO_3$ equivalent.

$$[Alk] = [HCO_3^-] + [CO_3^{2-}] + [OH^-] - [H^+] \tag{4.1}$$

Among the above anions, alkalinity due to bicarbonate ions is the major forms of alkalinity. This occurs in natural water and is known as natural alkalinity. Carbon dioxide is released by microbial action and is present in soil. The concentration of carbon dioxide is very high at topsoil compared to subsoil due to intense microbial actions. During rains, the water infiltrates into the soil and percolates through the soil layers. During the percolation, the water comes in contact with carbon dioxide and forms carbonic acid. The formation of carbonic acid creates a low pH condition in the soil layers. Thus, the minerals like calcium carbonate (from lime stone), magnesium carbonate (from dolomite) dissolve in this acidic solution as in Eqns 4.2 and 4.3; and form bicarbonates. These bicarbonates dissolve in water and cause alkalinity.

$$CaCO_3 + H_2CO_3 \longrightarrow Ca(HCO_3)_2 \quad (4.2)$$

$$MgCO_3 + H_2CO_3 \longrightarrow Mg(HCO_3)_2 \quad (4.3)$$

The dissolving of carbon dioxide in water generates carbonic acid and lowers the solution pH. This is the main reason behind the lower pH values of soft drinks. Farmers used to put lime in the field before the rainy season. Otherwise, the pH of soil decreases due to the formation of carbonic acid as explained earlier. The addition of lime neutralizes the carbonic acid formed during the rainy season.

Aeration is the best method to remove carbon dioxide from water. During the aeration, the pH of water increases with the removal of carbon dioxide. This is mainly due to the release of protons as in Eq. 4.4 [14]. Increase in water pH is a common phenomenon during the algal blooms. The pH of water may increase up to 10 in areas where algae grow at a rapid rate. This is also due to the consumption of carbon dioxide by the algae for their photosynthetic activity.

$$CO_2 + H_2O \rightleftharpoons H_2CO_3 \leftrightharpoons HCO_3^- + H^+ \quad (4.4)$$

The absence of bicarbonate ions is the common phenomena in boiler water. This is mainly due to the insoluble nature of carbon dioxide at elevated water temperatures. This results in an increased pH to the boiled water and cause shifting of ions from bicarbonate to carbonate and carbonate to hydroxide. Thus boiler water contains alkalinity due to carbonate and hydroxyl ions.

Alkalinity of water is an important parameter, which controls the efficiency of coagulants. Hydrolysable metal coagulants like alum, copperas, ferric chloride, sodium aluminate is efficient only in the presence of alkalinity, especially natural alkalinity. Addition of lime is required for the efficient operation of copperas. The alkalinity destroys the protons released by the coagulant and maintains the pH required for the effective formation of flocs.

Alkalinity concentration is also required for water softening process. Alkalinity, solution pH, soluble calcium and magnesium concentrations etc. are required to find the chemical dosages.

Among the various forms of alkalinity, the alkalinity due to hydroxyl ion (known as caustic alkalinity) is a major threat to our environment. The water containing caustic alkalinity should not be discharged into natural water bodies and sewers.

DETERMINATION OF ALKALINITY

Principle

The alkalinity of water is measured volumetrically using 0.02 N sulphuric acid as titrant. The titration curve for the alkalinity determination is shown in Fig. 4.1. The addition of acid to the water containing alkalinity reduces the solution pH. There are three specific points; A, B and C for the titration curve. Before the point 'A', all the hydroxyl ions react with protons and form water as in Eq. 4.5. The pH corresponding to this point is 10.5 and is known as carbonate equivalence point. At this point, all the caustic alkalinity (ie, alkalinity due to hydroxyl ions) is ceased by the reaction with protons. The second point, 'B' is known as bicarbonate equivalence point and the pH corresponding to this point is 8.3. Before this point, all the carbonate ions present in the water react with protons and form bicarbonate as in Eq. 4.6. The third point, 'C' is known as carbonic acid equivalence point and the solution pH corresponding to this point is 4.5. Before the carbonic acid equivalence point, all the bicarbonates present in water are converted to carbonic acid by the reaction with protons as in Eq. 4.7.

$$OH^- + H^+ \rightarrow H_2O \tag{4.5}$$

$$CO_3^{2-} + H^+ \rightarrow HCO_3^- \tag{4.6}$$

$$HCO_3^- + H^+ \rightarrow H_2CO_3 \tag{4.7}$$

From the titration curve, it can be seen that there is a sharp change in pH (point of inflection) only at two specific points, i.e. at bicarbonate and carbonic acid equivalence points. Therefore, only these changes can be identified in the presence of indicators. Phenolphthalein and mixed indicator are used as the indicator for this analysis. Mixed indicator is the combination of bromocresol green and methyl red. Phenolphthalein indicator point out the point of inflection at the bicarbonate equivalence point and indirectly gives the concentration hydroxyl and carbonate ions present in the water. Therefore, the alkalinity above pH 8.3 is known as Phenolphthalein alkalinity. At the same time mixed indicator specify the carbonic acid equivalence point and gives the concentration of bicarbonate ions. The alkalinity of water above the pH 4.5 is known as total alkalinity.

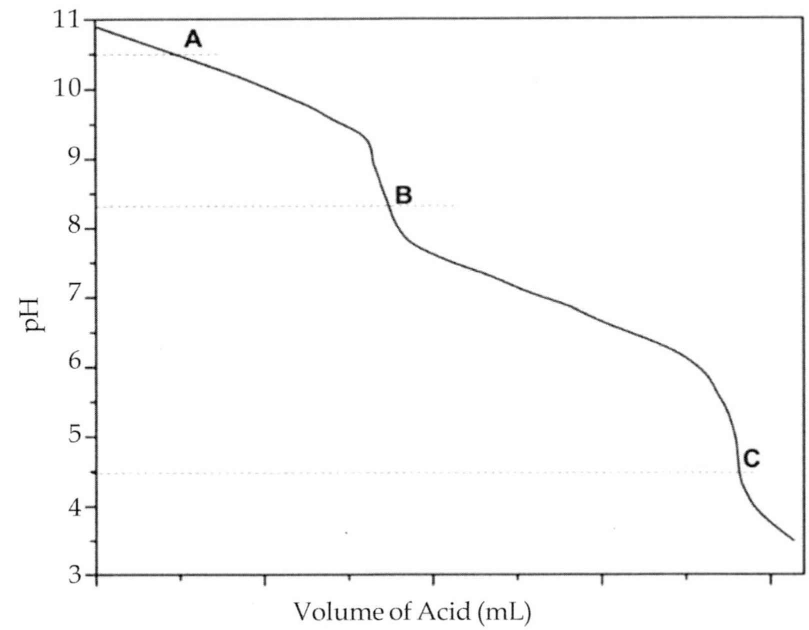

Fig. 4.1 Titration curve for the alkalinity determination

Reagent Preparation

Sulphuric acid of normality 0.02 N is used as the titrant for the analysis. This solution is prepared by diluting 0.53 mL of 98% pure H_2SO_4 to one litre. This solution can be standardized using 0.1 N sodium carbonate, in the presence of methyl orange indicator.

Mixed indicator and phenolphthalein are the indicators for the analysis. Mixed indicator is prepared by dissolving 100 mg bromocresol green sodium salt and 20 mg methyl red sodium salt in 100 mL distilled water; or 95% ethyl alcohol or isopropyl alcohol. Phenolphthalein indicator is prepared by dissolving 1 g of phenolphthalein in 200 mL of distilled water or ethyl alcohol.

Carbon dioxide free distilled water should be used for the preparation of all reagents, indicators etc because the dissolved carbon dioxide also imparts alkalinity to the water. This can be prepared by boiling the distilled water at least for 15 min and cooling at room temperature. The final pH of this water should be greater than 6 and its conductivity should be less than 2 μmhos/cm.

Standard Procedure and Chemistry behind the Analysis

1. Take 'V' mL of given sample in a clean Erlenmeyer flask.
2. If residual chlorine is present in the water sample, add sodium thiosulphate solution to remove the chlorine from water.

Chlorine is one of the major interferents for alkalinity analysis. The interfering reactions of chlorine during the alkalinity analysis are given in Eqns. 4.8 and 4.9. The interference of chlorine presented in the water can be ceased by the addition of sodium thiosulphate. Chlorine reacts with thiosulphate and reduces it to chloride as in Eq. 4.10.

$$Cl_2 + 2NaOH \rightarrow NaCl + NaOCl + H_2O \quad (4.8)$$

$$2Cl_2 + CaCO_3 + H_2O \rightarrow CaCl_2 + CO_2 + 2HClO \quad (4.9)$$

$$S_2O_3^{2-} + 4Cl_2 + H_2O \rightarrow 10H^+ + 2SO_4^{2-} + 8Cl^- \quad (4.10)$$

3. To this solution add two drops of phenolphthalein indicator. The solution becomes pink if the pH of water sample is greater than 8.3. Note down the initial burette reading (V_1).

Phenolphthalein is the most suitable indicator for showing the bicarbonate equivalence point. At pH 8.3, the phenolphthalein is colourless and has a pink colour at alkaline conditions. For example, consider the reaction between phenolphthalein and sodium hydroxide. Phenolphthalein has two phenol structures. The acidic part of these phenol groups react with sodium hydroxide and forms the sodium salt of phenolphthalein. This sodium salt of phenolphthalein imparts pink colour to the water.

4. Titrate this solution against standard sulphuric acid till the pink colour disappears. Note down the burette reading (V_2). The volume of sulphuric acid consumed for the phenolphthalein alkalinity (A) = V_2-V_1. Phenolphthalein alkalinity is calculated from the following equation.

Phenolphthalein Alkalinity (P)as mg/L of $CaCO_3$

$$= \frac{A \times \text{Normality of } H_2SO_4 \times 50 \times 1000}{\text{Volume of sample taken (mL)}} \quad (4.11)$$

5. To this solution, add two drops of mixed indicator.
6. Again titrate this solution against standard sulphuric acid, until the colour turns to pink. Note down the burette reading (V_3). The

volume of sulphuric acid consumed for the total alkalinity (B) = V_3-V_1. Total alkalinity is calculated from the following equation.

$$\text{Total Alkalinity (T)as mg/L of } CaCO_3 = \frac{B \times \text{Normality of } H_2SO_4 \times 50 \times 1000}{\text{Volume of sample taken (mL)}} \quad (4.12)$$

Both ingredients (methyl red and bromocresol green) are responsible for this colour change. The addition of mixed indicator gives a bluish shade to the water. After the determination of phenolphthalein alkalinity, the pH of water is 8.3. At pH above 5.4, bromocresol green imparts blue colour to the water. At this pH condition, bromocresol green is always in its dianionic form. At the same time, the methyl red imparts yellow colour to the water. Thus the initial colour of water is always a combination of these two colours. At the carbonic acid equivalent point, the colour of water changes to pink. At this pH condition, bromocresol green is always in its monoanionic form and has yellow colour. Methyl red imparts a red colour at this condition.

Even though phenolphthalein is present in the water, it will not give any colour change in between bicarbonate and carbonic acid equivalence points because phenolphthalein will not react with any acidic solution, neutral solution and water. This is mainly due to the presence of the phenolic groups present in phenolphthalein.

The amount of acid consumed for the titration is also dependent on the concentration of hydroxyl, carbonate and bicarbonate ions. Let, X be the amount of sulphuric acid consumed for the neutralization of hydroxyl ions. In a similar way, let Y and Z be the sulphuric acid concentration for the neutralization of carbonate and bicarbonate ions, respectively. If the solution contains only caustic alkalinity, it shows only the phenolphthalein end point. The water containing only carbonate ions show both end points, even though carbonate ions are present in water at elevated pH than bicarbonate equivalence point. Initially, the carbonate reacts with protons and form bicarbonate, which also cause alkalinity. Thus the same amount of bicarbonate formed from carbonate ions react with protons and form carbonic acid. The water with only carbonate ions will not show phenolphthalein end point. Such water will show only the total alkalinity end point. The volume of sulphuric acid consumption for various water samples is given in Table 4.1.

Table 4.1. Sulphuric acid consumption for various water samples

Constituents	Volume of Acid Consumed	
	pH>8.3	8.3>pH>4.5
Only OH^-	X	-
$OH^- + CO_3^{2-}$	X + Y	Y
Only CO_3^{2-}	Y	Y
$CO_3^{2-} + HCO_3^-$	Y	Y+Z
Only HCO_3^-	-	Z

Thus, it is very easy to calculate the concentrations of carbonate, bicarbonate and hydroxyl ions from the alkalinity analysis. By the comparison between the values of phenolphthalein alkalinity and total alkalinity values, the alkalinity due to these anions can be calculated. Hydroxide alkalinity is present in water when phenolphthalein alkalinity is more than total alkalinity or equal to total alkalinity. Carbonate alkalinity is present in water when phenolphthalein alkalinity is less than total alkalinity and is not zero. A zero value for phenolphthalein alkalinity or less than half the total alkalinity shows the presence of bicarbonate ions in water. Based on above concepts, the types of alkalinity can be calculated using the equations given in Table 4.2.

Table 4.2. Calculation of alkalinities based on P and T values

Result of Titration	Alkalinity due to (mg/L as $CaCO_3$)		
	OH^-	CO_3^{2-}	HCO_3^-
P = 0	0	0	T
P < 0.5 T	0	2P	T-2P
P = 0.5 T	0	2P	0
P > 0.5 T	2P-T	2(T-P)	0
P = T	T	0	0

The procedure described above for the determination of alkalinity concentration in water medium is shown in Fig. 4.2.

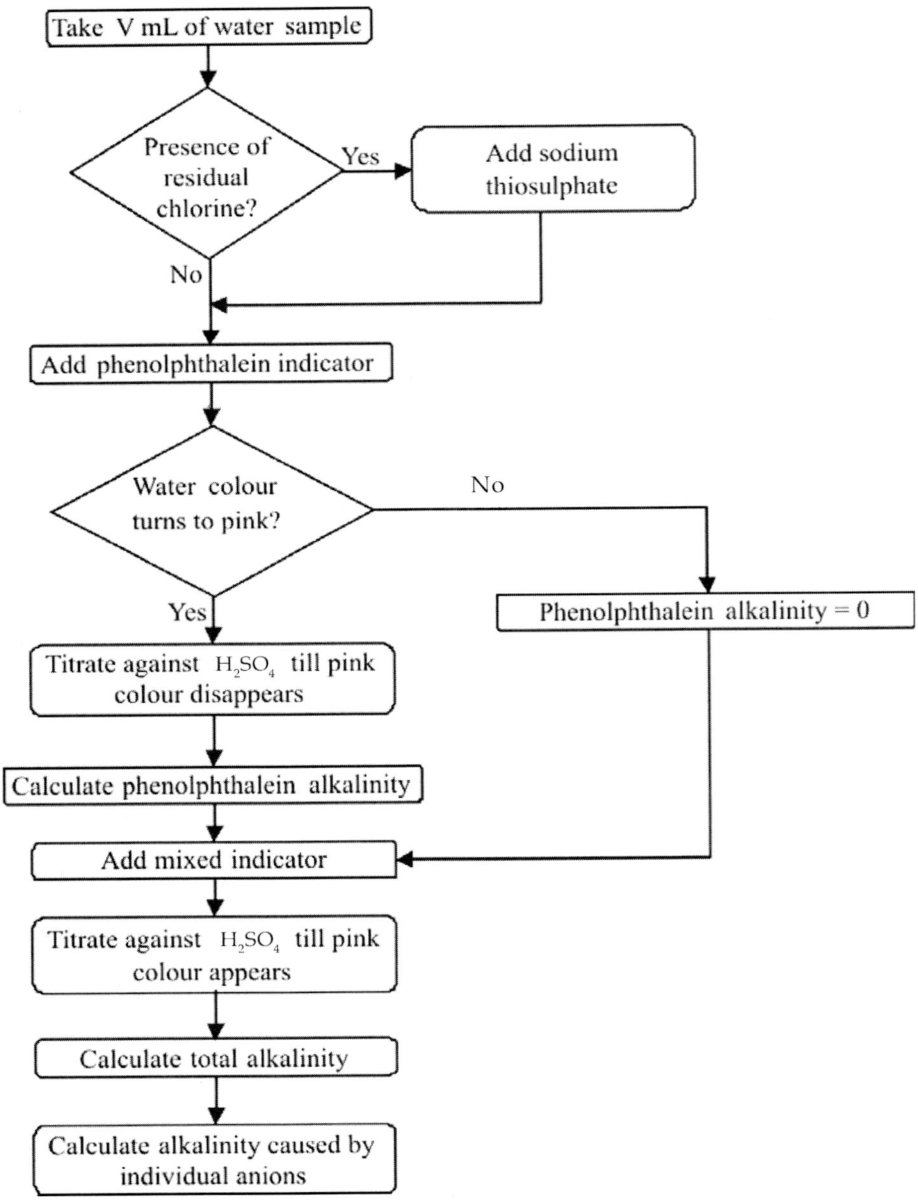

Fig. 4.2 Flowchart of alkalinity analysis procedure

TABULATION

Titration 1 (Phenolphthalein Alkalinity)

Burette solution: Sulphuric Acid

Indicator: Phenolphthalein

End Point: Disappearance of pink colour

No.	Volume of Sample (mL)	Burette Reading (mL)		Volume of H_2SO_4, A (mL)
		Initial	Final	

Titration 2 (Total Alkalinity)

Burette solution: Sulphuric Acid

Indicator: Mixed Indicator

End Point: Appearance of pink colour

No.	Volume of Sample (mL)	Burette Reading (mL)		Volume of H_2SO_4, B (mL)
		Initial	Final	

QUESTIONS

1. What you mean by natural alkalinity? Explain its formation.
2. What is the need for using carbon dioxide free distilled water for this volumetric analysis?

3. Alkalinity is an important parameter for chemical coagulation. Why? Explain with suitable chemical equations.
4. Explain the alkalinity titration curve.
5. With the help of neat sketch, explain the variation of carbonate, bicarbonate, carbonic acid concentrations with the changes in solution pH.
6. Explain the various forms of alkalinity in detail.
7. Why alkalinity is expressed in calcium carbonate scale?
8. Calculate the alkalinity as $CaCO_3$ of water that contains 185 mg/L of HCO_3^-, 245 mg/L of CO_3^{2-} and 10 mg/L of OH^-.
9. The alkalinity of water should be neither too low nor too high. Why?
10. Explain the interference of residual chlorine on alkalinity determination.
11. The pH of soft drinks is always acidic in nature. Why?
12. What is the need for applying lime on farm field?

OBJECTIVE QUESTIONS

1. The acid neutralizing capacity of water is termed as ....
2. Higher alkalinity cause .... taste to water
3. Water with low alkalinity is ....
4. Natural alkalinity is due to .... ions
5. The normality of sulphuric acid used for alkalinity determination is..................
6. The pH of water corresponding to carbonate equivalence point is ....
7. The pH of water corresponding to bicarbonate equivalence point is ....
8. The pH of water corresponding to carbonic acid equivalence point is ....
9. The alkalinity of water above the carbonate equivalence point is known as ....

10. Mixed indicator is a combination of .... and ....

11. .... is the most suitable indicator for showing bicarbonate equivalence point

12. The colour of phenolphthalein below pH 8.3 is.....................

13. The colour of dianionic form of bromocresol green is..........................

14. The colour of methyl red at carbonic acid equivalence point is ....

❏❏❏

5

Acidity

INTRODUCTION

The base neutralizing capacity of water is termed as acidity. Acidity of water is mainly due to the presence of strong mineral acids and weak acid like carbonic acid, acetic acid etc. Dissolved form of carbon dioxide is the major form, which is responsible for the acidity of water. Carbon dioxide-bicarbonates system is the major substances which give the buffering capacity to domestic wastewater, natural water, industrial wastewater etc. [14]. Natural processes are responsible for the acidity by dissolved carbon dioxide. As natural aeration, carbon dioxide also dissolves in water. Biological oxidation, both aerobic and anaerobic oxidation of organic matter generates carbon dioxide in water. Acidity due to mineral acids in water is caused by the effluents from the industrial wastes, especially from metallurgical industry. Salts of heavy metals also generate mineral acidity as a resultant of hydrolysis process.

Among all the components causing acidity, the acidity due to mineral acid, known as mineral acidity is a major threaten to environment. The water contain mineral acidity is unpalatable and affects the aquatic life. Thus, industrial wastewater containing mineral acidity should be neutralized before discharging into natural water bodies. The water containing acidity is not good for construction purposes. It corrodes steel and reduces the strength of concrete. Thus the water containing acidity more than 50 mg/L is not good for any reinforced cement concrete works. The presence of carbon dioxide in water also interfere water treatment processes, especially softening process.

DETERMINATION OF ACIDITY

Principle

The acidity of water is measured volumetrically using 0.02 N sodium hydroxide as titrant. Mineral acidity is present in water at pH below 4. All mineral acidity can be neutralized before the carbonic acid equivalence point. Thus, the presence of mineral acidity in water can be determined in the presence of methyl orange indicator. Methyl orange indicates the sharp changes in pH from 3.2 to 4.4. Methyl orange has red colour at pH below 4.4 and has yellow colour at pH 4.4. The half-way stage of methyl orange is at pH 3.7 and produces an orange colour at this pH [17]. Thus, this is the end point of this analysis, which shows the mineral acidity. Mineral acidity is also known as methyl orange acidity. Nowadays, bromphenol blue is using as the indicator for finding the methyl orange acidity. This indicator has a sharp colour change from yellow to purple at pH 3.7. The acidity due to carbon dioxide, known as carbon dioxide acidity, is determined by the presence of phenolphthalein indicator. Phenolphthalein indicator shows a sharp change in pH at bicarbonate equivalence point, i.e. at pH 8.3. The colour of water changes to pink at pH 8.3, due to the presence of phenolphthalein indicator. The colour change at pH 8.3 resembles the total acidity of water. Metacresol purple indicator can also be used for finding the total acidity of water. This indicator shows the end point as a colour change from yellow to purple.

Reagent Preparation

Sodium hydroxide of normality 0.02 N is the titrant for this analysis. This can be prepared by dissolving 0.8 g of sodium hydroxide in 1L of water. Methyl orange, bromphenol blue and metacresol purple indicators can be prepared by dissolving 100 g of their stuff or salt in 100 mL of distilled water. Carbon dioxide free distilled water should be used for all reagent preparation, dilution, etc.

Standard Procedure and Chemistry behind the Analysis

1. Take 'V' mL of given sample in a clean Erlenmeyer flask.
2. If residual chlorine is present in the water sample, add sodium thiosulphate solution to remove the chlorine from water.
3. To this solution add two drops of methyl orange or bromphenol blue indicator. Note down the initial burette reading (V_1).

4. Titrate this solution against standard sodium hydroxide solution until the end point. Note down the burette reading (V_2). The volume of sodium hydroxide consumed for the mineral acidity (A) = V_2-V_1. Then, mineral acidity is calculated from the following equation.

$$\text{Mineral Acidity as mg/L of } CaCO_3 = \frac{A \times \text{Normality of NaOH} \times 50 \times 1000}{\text{Volume of sample taken (mL)}} \quad (5.1)$$

5. To this solution add two drops of phenolphthalein or metacresol purple indicator.

6. Again titrate this solution against standard sodium hydroxide, until the end point. Note down the burette reading (V_3). The volume of NaOH consumed for the total acidity, B = V_3-V_1. Total alkalinity is calculated from the following equation.

$$\text{Total Acidity as mg/L of } CaCO_3 = \frac{B \times \text{Normality of NaOH} \times 50 \times 1000}{\text{Volume of sample taken (mL)}} \quad (5.2)$$

The procedure described above also shown in Fig. 5.1 in the form of a flowchart.

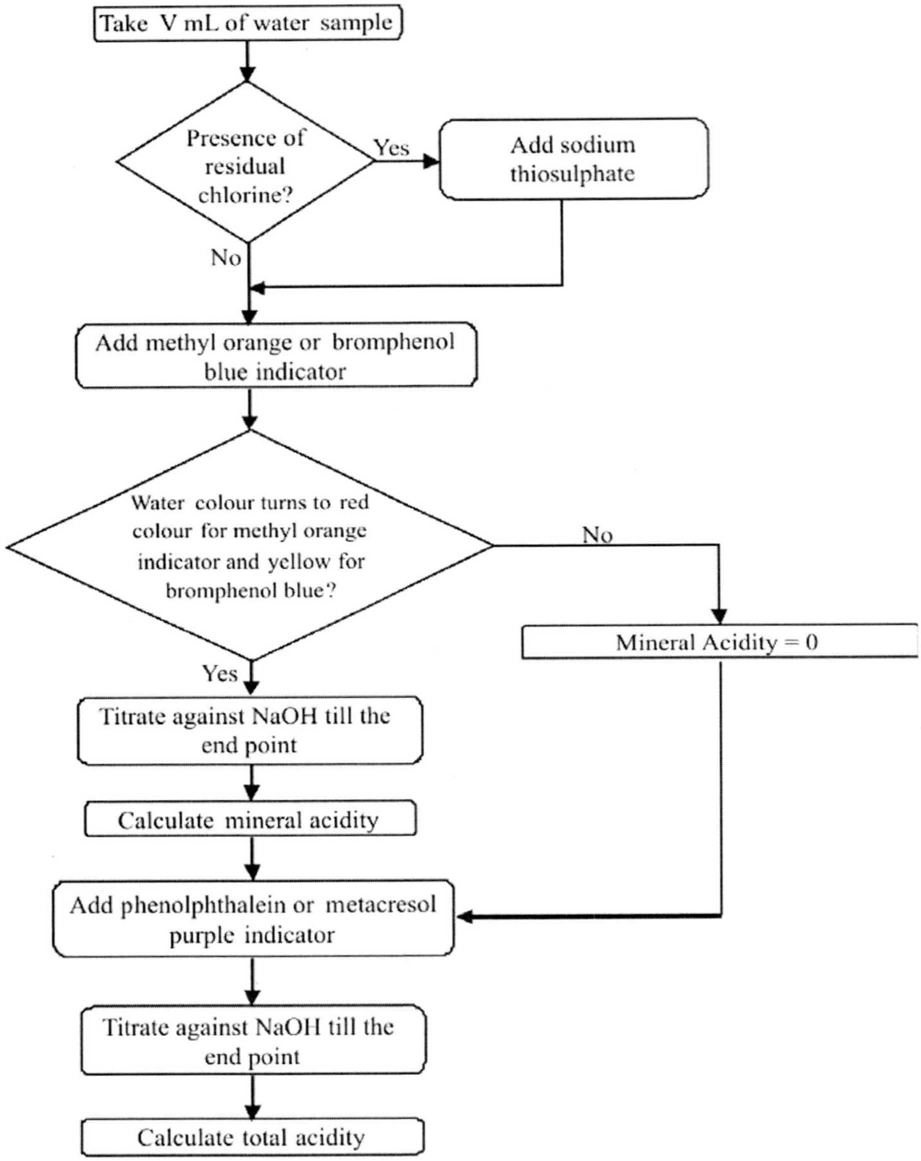

Fig. 5.1 Flowchart of acidity analysis procedure

TABULATION

Titration 1: Mineral Acidity

Burette solution: Sodium hydroxide

Indicator: Methyl orange or bromphenol blue

End Point: Appearance of orange colour for methyl orange indicator; Yellow to purple for bromphenol blue indicator

No.	Volume of Sample (mL)	Burette Reading (mL)		Volume of NaOH A (mL)
		Initial	Final	

Titration 2: Total Acidity

Burette solution: Sodium Hydroxide

Indicator: Phenolphthalein or Metacresol purple

End Point: Appearance of pink (Phenolphthalein) or purple (Metacresol purple).

No.	Volume of Sample (mL)	Burette Reading (mL)		Volume of NaOH B (mL)
		Initial	Final	

QUESTIONS

1. What are the ill-effects of acidity?
2. Explain the principle of acidity determination.

3. What are the major sources of acidity in natural water?
4. What pH ranges is used to measure mineral acidity and total acidity?
5. What are the applications of acidity data in environmental engineering?

OBJECTIVE QUESTIONS

1. The base neutralizing capacity of water is termed as....................
2. Mineral acidity is present in water at pH ....
3. The half-way stage of methyl orange is at pH ....
4. The coloured of methyl orange at pH 3.7 is....................
5. ....is also known as methyl orange acidity.
6. Nowadays, instead of methyl orange ....is using as the indicator for finding the methyl orange acidity.
7. Bromphenol blue indicator has a sharp colour change from .... to ....at pH 3.7.
8. ....indicator can also use for finding the total acidity of water.

❑❑❑

6

Hardness

INTRODUCTION

The characteristic of water which does not form sufficient lather by the application of soap is commonly known as hardness. The hardness of water is mainly due to the cations like calcium, magnesium, strontium, iron, manganese etc. and anions like bicarbonate, carbonate, chloride, sulphate, nitrate, silicate etc. Compared to the concentration of other cations, the amount of calcium and magnesium is very high in the natural water. Thus, hardness of water is mainly due to the calcium and magnesium ions present in water and its associated anions. Hardness of water is always expresses in calcium carbonate equivalent scale. The total hardness of water is computed from the Eq. 6.1. Thus, total hardness is the sum of calcium hardness and magnesium hardness. Depends on the degree of hardness, the water is classified as soft water (hardness below 75 mg/L), moderately hard (hardness in between 75 and 150 mg/L), hard water (hardness in between 150 and 300 mg/L) and very hard (hardness greater than 300 mg/L) [14].

$$\text{Total Hardness (mg/L as } CaCO_3\text{)} =$$

$$[Ca^{2+} \text{in mg/L} \; \frac{\text{Equivalent Weight of } CaCO_3}{\text{Equivalent Weight of } Ca^{2+}} + [Mg^{2+} \text{in mg/L} \; \frac{\text{Equivalent Weight of } CaCO_3}{\text{Equivalent Weight of } Mg^{2+}}]$$

$$= \left[Ca^{2+} \text{in mg/L} \; \frac{50}{20}\right] + \left[Mg^{2+} \text{in mg/L} \; \frac{50}{12}\right] \tag{6.1}$$

Hardness of water is divided in to two: carbonate hardness and non-carbonate hardness. Carbonate hardness is the hardness of water caused by carbonates and bicarbonates of calcium and magnesium ions. Carbonate hardness also cause alkalinity to the water. Thus, carbonate hardness of water is equal to total hardness or alkalinity, whichever is less [11]. Non-carbonate hardness is the hardness of water caused by the anions other than carbonate and bicarbonate. Thus, non-carbonate hardness is caused mainly by the chlorides and sulphates of calcium and magnesium ions. Non-carbonate hardness of water is equal to total hardness minus carbonate hardness [14].

Hardness of water can also be divided in to temporary hardness and permanent hardness. Temporary hardness of water can be removed by simple boiling and is caused by carbonates and bicarbonates of calcium and magnesium. Thus, temporary hardness is another name of carbonate hardness. The main disadvantage of temporary hardness is the scale formation. For example, the heating of calcium bicarbonate precipitates as calcium carbonate (Eq. 6.2). This precipitate sticks on the sides of utensils, pipes etc. (formation of scale). Scale formed inside the utensils act as an insulator and increases the energy cost. The scale formation on inner side of pipes reduces the corrosion. At the same time, it reduces the carrying capacity of pipelines.

$$Ca(HCO_3)_2 \rightarrow CaCO_3 + H_2O + CO_2 \qquad (6.2)$$

Permanent hardness of water cannot be removed by simple boiling and requires special treatment such as lime soda process, caustic soda process etc. Permanent hardness of water is caused by non-carbonate hardness.

Insufficient lather formation by the application of soap is another problem with hard water, as explained earlier. This is mainly due to the complex formation by the reaction between divalent cations and the soap. The divalent cations present in hard water reacts with soap and form the precipitate, known as scum. Thus, the lather will not form with the application of soap, until all the divalent cations present in water ceased.

Insufficient lather is also not formed in the case of sea water. This interference is mainly due to the higher concentration of sodium ions, even though which is not responsible for hardness. The higher sodium concentration in water responsible for common ion effect and reduces the foam formation. This is known as pseudo-hardness.

Even though hardness of water is responsible for scale formation and insignificant lather formation, the drinking water should contain some amount of hardness. Absolutely soft water is corrosive and tasteless. Soft water is also responsible for cardio vascular diseases. Due to the presence of calcium ions, hard water is helpful for the growth of children. But, higher amount of hardness impart bitter taste to water, as in the case of alkalinity. Laxative effects associated with magnesium sulphate are also a problem of hard water.

DETERMINATION OF TOTAL HARDNESS

Principle

The total hardness of water is found by titration the water sample against Ethylenediamine tetra-acetic acid (EDTA) or its sodium salts. Divalent cations causing hardness form a chelated soluble complex with EDTA as in Eq. 6.3. Eriochrome black T (EBT) is the indicator for this volumetric analysis. Calcium and magnesium ions present in water forms a complex at a pH of 10 ± 0.1 as in Eq. 6.4. and have a wine red colour. Between these complexes, the complex between divalent cations and EDTA is more stable. Initially, the divalent cations present in water forms the complex with EBT. As the titration with EDTA continues, EDTA disrupts the EBT complex and forms the complex with divalent cations. This releases the EBT indicator and shows a sharp end point from wine red to blue colour.

$$M^{2+}EDTA \rightarrow [M.EDTA]_{complex} \tag{6.3}$$

$$M^{2+}EBT \rightarrow [M.EBT]_{complex} \tag{6.4}$$

Reagent Preparation

Standard EDTA solution of normality 0.02 N is commonly used as the titrant for this analysis. It can be prepared by dissolving 3.723 g analytical reagent-grade di-sodium EDTA, also called (ethylenedinitrilo) tetraacetic acid disodium salt in 1L of distilled water. EBT indicator can be prepared by dissolving 0.5 g EBT in 100 mL distilled water. The amount of indicator to be added for the titration should be 2 drops per 50 mL of water sample.

Ammonia buffer is used in this titration to maintain the alkaline pH. This buffer can be prepared by dissolving 16.9 g ammonium chloride (NH_4Cl) in 143 mL concentrated ammonium hydroxide (NH_4OH). To this solution add 1.25 g magnesium salt of EDTA and dilute to 250 mL with distilled water. Magnesium salt of EDTA is a commercially available chemical. In the absence of this salt, the buffer can prepare using the disodium salt of EDTA. To 50 mL of distilled water, add 1.179 g disodium salt of EDTA and 780 mg magnesium sulfate ($MgSO_4.7H_2O$) or 644 mg magnesium chloride ($MgCl_2.6H_2O$). Mix thoroughly until the salts are dissolved. Add this solution and 16.9 g NH_4Cl to 143 mL concentrated NH_4OH. Dilute the solution to 250 mL with distilled water.

Standard Procedure and Chemistry behind the Analysis

1. To a constant volume of water sample taken in an Erlenmeyer flask, add 2 mL of ammonia buffer solution.

The addition of ammonium buffer maintains the water pH near 10. This pH is required for the complex formation reaction between EBT and divalent cations.

2. To this solution add two drops of EBT indicator solution.

The solution colour change to wine red is the indication of hardness present in the water. The divalent cations form the complex with EBT and it has wine red colour. For soft water, colour doesn't change to wine red after the addition EBT.

3. Titrate this solution against standard EDTA solution until the end point reaches (appearance of blue colour).

The appearance of blue colour indicates the release of EBT to the solution. The wine red coloured EBT-divalent cations have been broken by the addition of EDTA and form EDTA-divalent cation complex.

Note: Generally a violet colour will appear before the appearance of blue colour. This is not the end point. The end point of this analysis is a sharp colour changes from this violet (or from wine-red) to steel blue colour.

4. Note down the volume of EDTA added (V) and calculate the total hardness as:

$$\text{Total Hardness as mg/L } CaCO_3 = \frac{V \times 1000}{\text{Volume of sample taken (mL)}} \qquad (6.5)$$

The total hardness analysis procedure is also shown in Fig. 6.1.

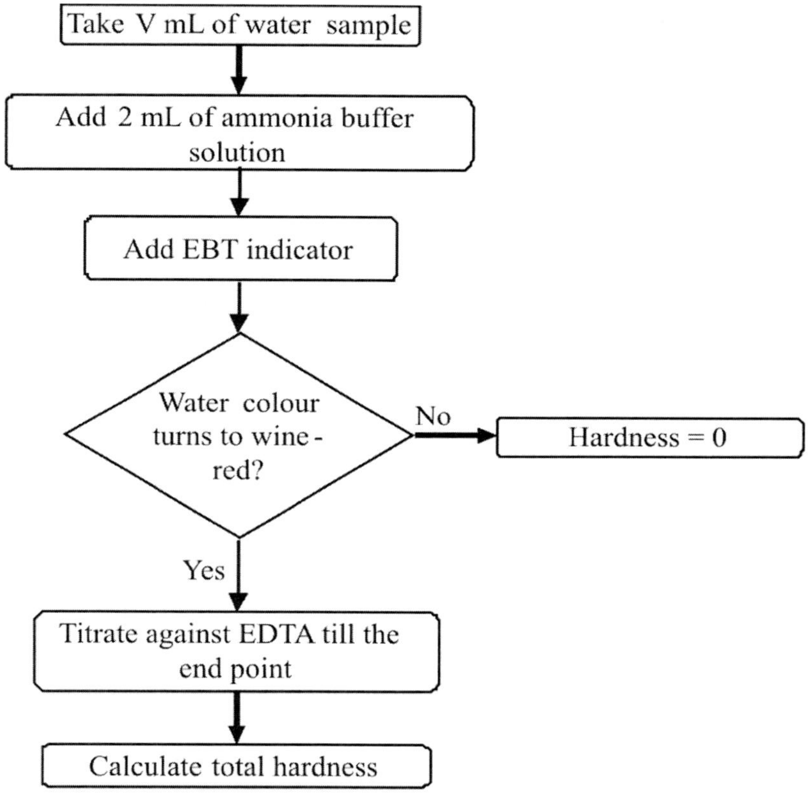

Fig. 6.1 Flowchart of hardness analysis procedure

TABULATION

Burette solution: EDTA

Indicator: EBT

End Point: Wine red to blue

No.	Volume of Sample (mL)	Burette Reading (mL)		Volume of EDTA (mL)
		Initial	Final	

DETERMINATION OF CALCIUM HARDNESS

Principle

Calcium and magnesium ion forms chelated soluble complex with EDTA. The complex formation between magnesium and EDTA can be suppressed by conducting the volumetric analysis at elevated pH conditions. At pH above 11, all the magnesium ions present in water form magnesium hydroxide precipitate. Thus at highly alkaline conditions, only calcium forms complex with EDTA. Ammonium purpurate is the indicator for this analysis. Ammonium purpurate forms complex with calcium and imparts a pink colour. If EDTA is then added as a titrant, all free calcium ions form complex with EDTA. Finally, the EDTA disrupts the pink complex (Ca. ammonium purpurate) because it is capable of forming more stable complex with the calcium ions. This action frees the ammonium purpurate indicator, and pink colour changes to a distinct purple colour, heralding the end point of titration.

Reagent Preparation

Standard EDTA (0.02 N) is the titrant for this analysis and its preparation procedure is explained above. Ammonium purpurate (commonly known as murexide) indicator is prepared by mixing 0.5 g of ammonium purpurate and 100 g NaCl. Use this mix as dry powder. Sodium hydroxide of normality 1N is used to for increasing the water pH and is prepared by dissolving 40 g NaOH in 1 litre distilled water.

Standard Procedure and Chemistry behind the Analysis

1. Take a constant volume of water sample in an Erlenmeyer flask.
2. To the sample, add 1 mL of 1 N NaOH solution and a pinch of ammonium purpurate powder.

The addition of sodium hydroxide increases the pH above 11. At this pH condition, all the magnesium ions present in water precipitates as magnesium hydroxide. The colour change of water to pink with the addition of ammonium purpurate indicates the presence of calcium hardness in water. Calcium ions form a complex with ammonium purpurate and have the pink colour.

3. Titrate this against standard EDTA solution until the appearance of purple colour.

EDTA breaks the calcium-ammonium purpurate complex and form complex with calcium. This reaction release ammonium purpurate into the solution. Thus, the purple colour at the endpoint indicates the presence of ammonium purpurate in water.

4. Note down the volume of EDTA added (V) and calculate the calcium hardness as:

$$\text{Calcium Hardness as mg/L } CaCO_3 = \frac{V \times 1000}{\text{Volume of sample taken (mL)}} \quad (6.6)$$

Calcium hardness in terms of calcium present in water mg/L

$$= \text{Calcium Hardness as mg/L } CaCO_3 \times \frac{\text{Eq.Wt.of calcium}}{\text{Eq.Wt.of } CaCO_3}$$

$$= \text{Calcium Hardness as mg/L } CaCO_3 \times \frac{20}{50} \quad (6.7)$$

Fig. 6.2 Also shows the calcium hardness determination procedure

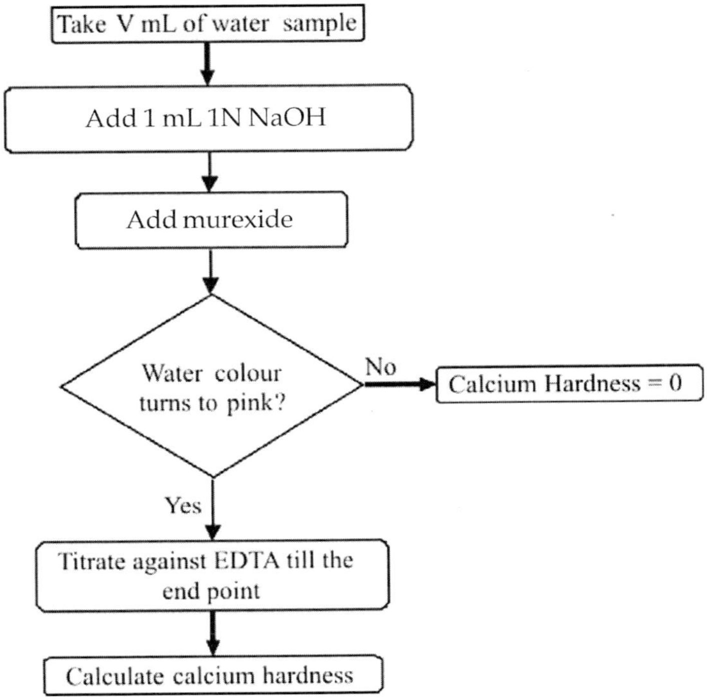

Fig. 6.2 Flowchart of calcium hardness analysis procedure

TABULATION

Burette solution: EDTA

Indicator: Ammonium purpurate

End Point: Pink to purple

No.	Volume of Sample (mL)	Burette Reading (mL)		Volume of EDTA (mL)
		Initial	Final	

QUESTIONS

1. Define hardness of water. What are the major anions and cations, which are responsible for hardness?
2. What are the differences between scum and scale?
3. What are the ill-effects of hardness?
4. Classify water based on degree of hardness.
5. Can you determine temporary hardness and permanent hardness separately? If yes, how?
6. Explain the classification of water hardness. Explain its properties and relationships.
7. What is pseudo-hardness?

OBJECTIVE QUESTIONS

1. The precipitate formed by the reaction between divalent cations and soap is known as ....
2. EDTA is used as a .... agent
3. Magnesium sulphate cause .... effect
4. The other name of ammonium purpurate is...................
5. Temporary hardness is also known as ....

6. .... Hardness is responsible for scale formation
7. Insufficient lather formation in the presence of higher sodium concentration is known as ....
8. Soft water is also responsible for ....diseases
9. The colour of divalent cation –EBT complex is ....
10. Ammonium buffer is the combination of .... and ....

❑❑❑

7

Dissolved Oxygen

INTRODUCTION

Dissolved oxygen (DO) concentration in water indirectly indicates the water purity, especially in the case of surface water. High concentration of dissolved oxygen resembles high purity of water. Therefore, dissolved oxygen concentration is considered as a major parameter for monitoring the surface water quality.

The amount of dissolved oxygen in water depends on various physico-chemical and biological parameters. The major source of DO in water is by natural aeration. The atmospheric oxygen diffuses into the water and enhances the DO concentration. This diffusion rate depends mainly on the temperature of water and follows Henry's law. Another source of DO in water is the algae present in water. Algae release oxygen as a by-product of photosynthesis. Algal bloom in water may shoot-up the DO concentration to a high level, called super saturation. The maximum amount of DO in water, known as saturation DO level (DO_{sat}) is dependent on water temperature (as per Henry's law). The value of DO_{sat} as a function of temperature is given as [18]:

$$DO_{sat} = 14.652 - 0.41022T - 0.0079910T^2 - 0.000077774T^3 \qquad (7.1)$$

where DO_{sat} is the DO saturation concentration in mg/L and T is the water temperature in °C.

With the increase in DO concentration in water by natural aeration and algal photosynthesis, the DO is consumed by various other sources. Algae present in the water itself use DO during their respiration period. Bacteria use DO for their growth and for the degradation of organic pollutants. Aquatic animals also use DO for their survival. Thus the amount of DO in water depends on temperature, rate of biological activities, amount of organic impurities, amount of dissolved solids, etc.

Overall, the rate of re-aeration is always higher than rate of de-aeration. This is the main reason behind the natural purification capacity of streams and rivers. The introduction of pollutants from industries or from other sources depletes the DO content in the water. This decreased DO is mainly used for the degradation of pollutants. The organic pollutants are degraded by biological activities. The zone of river or stream at which the decomposition of pollutants occurs is known as zone of active decomposition or zone of degradation. Presence of carp, mosquito and black-fly larva etc. in water indicates the degradation zone of river. If the pollutant concentration is very high, the DO content in water gets reduced to a very less value, even to zero and cause anaerobic conditions in the stream or river. This zone is known a septic zone. The absence of fish is a good indication of septic zone. Apart from this, this zone contains worms, larva of mosquito and midge etc. Due to the high rate of re-aeration, the DO content in the water increases to a desired level. During this, the pollutants are degraded in a rapid manner. The zone of river showing the increasing DO level is known as recovery zone. At last, the DO level in the water reaches to its initial value and the zone is known as clear zone. At this zone, the concentration of pollutant is minimal.

The higher concentration of DO gives good taste to water. Higher DO concentrations in water shoot up the aerobic biological activities in water. At the same time higher DO concentration is responsible for the corrosion of steel, iron etc. Survival of aquatic life is also dependent on DO concentration. Most of the aquatic animals require high oxygen content for their survival. For example, species like summer flounder, striped bass, hard clams etc. require DO of at least 5 mg/L for their survival. But, the species like crabs, worms, etc. will survive at low DO concentrations. Considering all the species in the water, a minimum DO of 4 mg/L is considered as the standard DO concentration for the survival of aquatic life.

DETERMINATION OF DO

Principle

Winkler method is commonly used to determine the DO concentration present in the water sample. The divalent manganese salt added externally into the water reacts with dissolved oxygen present in the water and forms manganese dioxide precipitate (Eq. 7.2). Manganese hydroxide formed by the hydrolysis of divalent manganese ions also undergoes oxidation reaction and forms manganese dioxide (Eq. 7.3) or hydrated manganese oxide (Eq. 7.4). All the dissolved oxygen present in the water is consumed by the addition of divalent manganese ions and precipitates as brown $MnO(OH)_2$ or MnO_2. Since, all the oxygen is consumed by the addition of divalent manganese ions, this reaction is known as oxygen fixation reaction.

$$Mn^{2+} + 2OH^- + 0.5O_2 \rightarrow MnO_2 + H_2O \qquad (7.2)$$

$$Mn(OH)_2 + 0.5O_2 \rightarrow MnO_2 + H_2O \qquad (7.3)$$

$$2Mn(OH)_2 + O_2 \rightarrow 2MnO(O\,H)_2 \text{ (Brown Precipitate)} \qquad (7.4)$$

Further addition of alkali iodide and acid to this solution liberates iodine as in Eqns. 7.5 and 7.6. The amount of liberated iodine can be measured using standard thiosulphate. Thiosulphate reduce this iodine to iodide as in Eq. 7.7. Starch indicates the end point of this reaction. The amount of iodine is equivalent to the amount of oxygen dissolved in the water sample.

$$MnO_2 + 2I^- + 4H^+ \rightarrow Mn^{2+} + I_2 + H_2O \qquad (7.5)$$

$$MnO(OH)_2 + 2KI + H_2O \rightarrow Mn(OH)_2 + 2KOH + I_2 \qquad (7.6)$$

$$2S_2O_3^{2-} + I_2 \rightarrow S_4\,O_6^{2-} + 2I^- \qquad (7.7)$$

Reagent Preparation

Sodium thiosulphate having 0.02 N is the titrant used for the DO analysis. This solution can be prepared by dissolving 6.205 g of sodium thiosulphate ($Na_2S_2O_3$. $5H_2O$) in distilled water, followed by the addition of 1.5 mL 6N sodium hydroxide or 0.4 g solid sodium hydroxide. Dilute this solution to 1 L and store in a well tightened bottle.

Starch indicator, used commonly to detect the iodine presence in water and it can be prepared by dissolving 2 g of laboratory-grade soluble starch in 100 mL of hot distilled water. Freshly prepared starch is required for the analysis. Otherwise, the addition of 0.2 g salicylic acid during the preparation of starch indicator preserves the indicator.

Manganous sulphate solution can prepare by dissolving 400 g of dihydrated manganese sulphate ($MnSO_4.2H_2O$), or 480 g of tetrahydrated manganese sulphate ($MnSO_4.4H_2O$) or 364 g of monohydrated manganese sulphate ($MnSO_4.H_2O$) in freshly boiled and cooled distilled water. Make-up the solution volume to 1L by adding sufficient amount of distilled water.

Alkali-iodide-azide reagent can be prepared by dissolving 500 g of sodium hydroxide and 135 g of sodium iodide (NaI) in distilled water. Instead of sodium salts, 700 g of potassium hydroxide and 150 g of potassium iodide (KI) can be used. Make the final volume of the reagent as 1L by adding distilled water. Dissolve 10 g of sodium azide (NaN_3) in 40 mL distilled water and add to the alkali-iodide solution.

Standard Procedure and Chemistry behind the Analysis

1. Fill the water sample in a BOD bottle and put the stopper to remove the excess water. Remove the stopper and add 2 mL of manganous sulphate solution and 2 mL of alkali-iodide azide reagent at the bottom of BOD bottle or at least well below the surface of the liquid. Put the stopper carefully to exclude the air bubbles.

The addition of manganous sulphate results in oxygen fixation reaction. But, the oxygen fixation reaction is a slow process. Thus, all the oxygen may not take place the fixation reaction. Therefore, vigorous shaking of the sample after the manganous sulphate addition is required for fixing the entire DO present in the water. This can be achieved by inverting the bottle at least 15 times after the addition of reagents.

Oxygen fixation reaction is an important step in Winkler method. After the fixation reaction, the sample can store up to 8 hrs. Thus, this step is very useful to find the exact concentration of DO while monitoring the water quality of a river/stream stretch. In this case, several samples should be analyzed at a time. Also, the samples should be analyzed within a short period. Otherwise, dissolved oxygen concentrations in the sample may alter due to the reactions in the BOD bottle, or other changes in physico-chemical parameters. This change in DO value can be regulated by adding manganous sulphate and alkali-iodide azide reagent. The addition of manganous sulphate fix the DO present in the water and the sample can be analyzed even after 8 hrs.

Sometimes, instead of brown precipitate, white precipitate may form after the addition of manganous sulphate. This indicates the absence of DO in the water sample. In the absence of oxygen, divalent manganese ions precipitate as $Mn(OH)_2$ (Eq. 7.8) and has a white colour.

$$Mn^{2+} + 2OH^- \rightarrow Mn(OH)_2 \text{ (White Precipitate)} \qquad (7.8)$$

2. After the sufficient mixing, allow the brown precipitate to settle down.

3. Add 1 mL of concentrated sulphuric acid to the bottle after the settling of precipitates. Re-stopper the bottle and shake well, until the entire precipitates dissolves.

Since, sulphuric acid is highly hygroscopic, special care should be taken while adding the acid to water. Don't add the acid directly to the water sample. This acid should add through the neck of the bottle.

The addition of sulphuric acid generates an acidic medium required for the iodine liberation reaction. Sometimes, 1 mL of the acid is not sufficient for dissolving entire brown precipitate. In that case add additional sulphuric acid, say 1 mL to dissolve the precipitates. All the manganese precipitates should dissolve before the volumetric analysis. Because, in Winkler method, DO concentration is determined by measuring this liberated iodine concentration and the amount of iodine liberated is directly proportional to the amount of precipitates dissolved in the sample.

Shaking of bottle at least for 10 to 15 times after the addition of acid is necessary for the formation of solution by dissolving the precipitate and for the effective liberation of iodine. Insoluble iodine gas may reacts with iodide and form more soluble tri-iodide as in Eq. 7.9. The generation of tri-iodide prevents the escape of iodine gas. With the effective mixing, the iodine distributes uniformly throughout the sample and generates tri-iodide.

$$I_2 + I^- \rightarrow I_3^- \qquad (7.9)$$

4. Transfer 203 mL of the solution from the BOD bottle to an Erlenmeyer flask.

As per the standard procedure, 200 mL of sample is required for the analysis. The additional 3 mL sample volume for the analysis is the correction volume for the amount of manganous sulphate and alkali-iodide-azide reagent added in the previous step. The addition of 2 mL manganous sulphate and 2 mL alkali-iodide-azide reagent reduced the effective sample volume from 300 mL to 296 mL. Therefore, the equivalent sample volume required for 200 mL sample is

$$= \frac{200 \times 300}{300 - 4} = 203\,\text{mL}$$

5. Titrate this sample against 0.025 N sodium thiosulphate solution until a pale yellow colour reaches. Then add 1 mL of starch indicator. The addition of starch produces a blue colour to the solution and continue the titration till the disappearance of blue colour.

Even though iodine gas liberated by the addition of acid, iodine gas doesn't impart any colour to the water sample. The colour of water with the addition of sulphuric acid is mainly due to the formation of tri-iodide by the reaction between iodine gas and iodide as in Eq. 7.9. Tri-iodide imparts different colour to the water with its concentrations. The colour of water varies from light to brown with the concentration of tri-iodide. Sometimes, the colour of water may change to violet at high tri-iodide concentration.

The concentration of tri-iodide can estimated in the presence of starch indicator. Starch forms blue colour complex with tri-iodide. With the addition of thiosulphate to the water containing tri-iodide starch complex, thiosulphate reacts with tri-iodide and forms iodide as in Eq. 7.10. Thus, the bond between tri-iodide and starch breaks at the end point and results in the disappearance of blue colour. During this analysis, a special precaution should be taken for the addition of starch. Starch indicator should be added after decreasing the concentration of tri-iodide by adding thiosulphate (i.e. making the colour of solution to pale yellow by adding thiosulphate). Otherwise, higher tri-iodide content destroys starch and reduces the complex formation reaction. Also, the complex formation and desorption reaction between tri-iodide and starch is very fast and reversible. At higher tri-iodide concentration, the bonding between starch and tri-iodide is too strong. This retards the desorption process and makes it difficult to identify the end point easily [19].

$$2S_2O_3^{2-} + I_3^- \rightarrow S_4O_6^{2-} + 3I^- \qquad (7.10)$$

There are several interferents for the DO analysis by Winkler method. Reducing agents like Fe^{2+}, SO_3^{2-}, S^{2-} and polythionate reduce iodine to iodide. Thus, the presence of these ions produces results, which is very less than the actual value. Apart from this, oxidizing agents like nitrate, Fe^{3+} oxidize iodide to iodine and produce results that are too high.

The nitrate ion is one of the most frequent and dangerous interferences encountered in the DO determination [14]. Nitrate ion oxidizes iodide to iodine under acidic conditions (Eq. 7.11). The generated N_2O_2 reacts with oxygen and forms nitrate again in the solution (Eq. 7.12). Thus, the presence of nitrate in water sample establishes cyclic interference reactions during the analysis and produce high results, far in excess of amounts that could be expected. After the completion of thiosulphate-iodine reaction (disappearance of blue colour after starch addition), nitrate ions reacts with the iodide and forms iodine. This will again gives blue colour to the solution. Thus, it is impossible to get a permanent end point in the presence of nitrate [14].

Sodium azide is the most efficient substance which can overcome the nitrate interference. Thus, in azide modified Winkler method, azide is incorporate in alkali iodide reagent. In an acidic medium, azide converts nitrite to N_2 and N_2O (Eqns. 7.13 and 7.14) and thus, suppresses the interference caused by nitrite.

$$2NO_2^- + 2I^- + 4H^+ \rightarrow I_2 + H_2O + N_2O_2 \tag{7.11}$$

$$N_2O_2 + 0.5O_2 + H_2O \rightarrow 2NO_2^- + 2H^+ \tag{7.12}$$

$$NaN_3 + H^+ \rightarrow HN_3 + Na^+ \tag{7.13}$$

$$HN_3 + H^+ + NO_2^- \rightarrow N_2 + N_2O + H_2O \tag{7.14}$$

6. Document the amount of sodium thiosulphate consumed for the entire titration. The normality of sodium thiosulphate and sample volume for the analysis are selected in such a way that the consumed sodium thiosulphate volume directly gives the dissolved oxygen concentration in mg/L. The calculation for DO concentration is as follows.

$$\text{Concentration of oxygen} = \frac{V \times \text{Normality of } Na_2S_2O_3 \times \text{Eq.Wt.of } O_2 \times 1000}{\text{Volume of sample taken (mL)}}$$

$$= \frac{V \times 0.025 \times 8 \times 1000}{200} = V \text{ mg/L} \tag{7.15}$$

The procedure described above for the DO determination from the aqueous solution is also shown in Fig. 7.1.

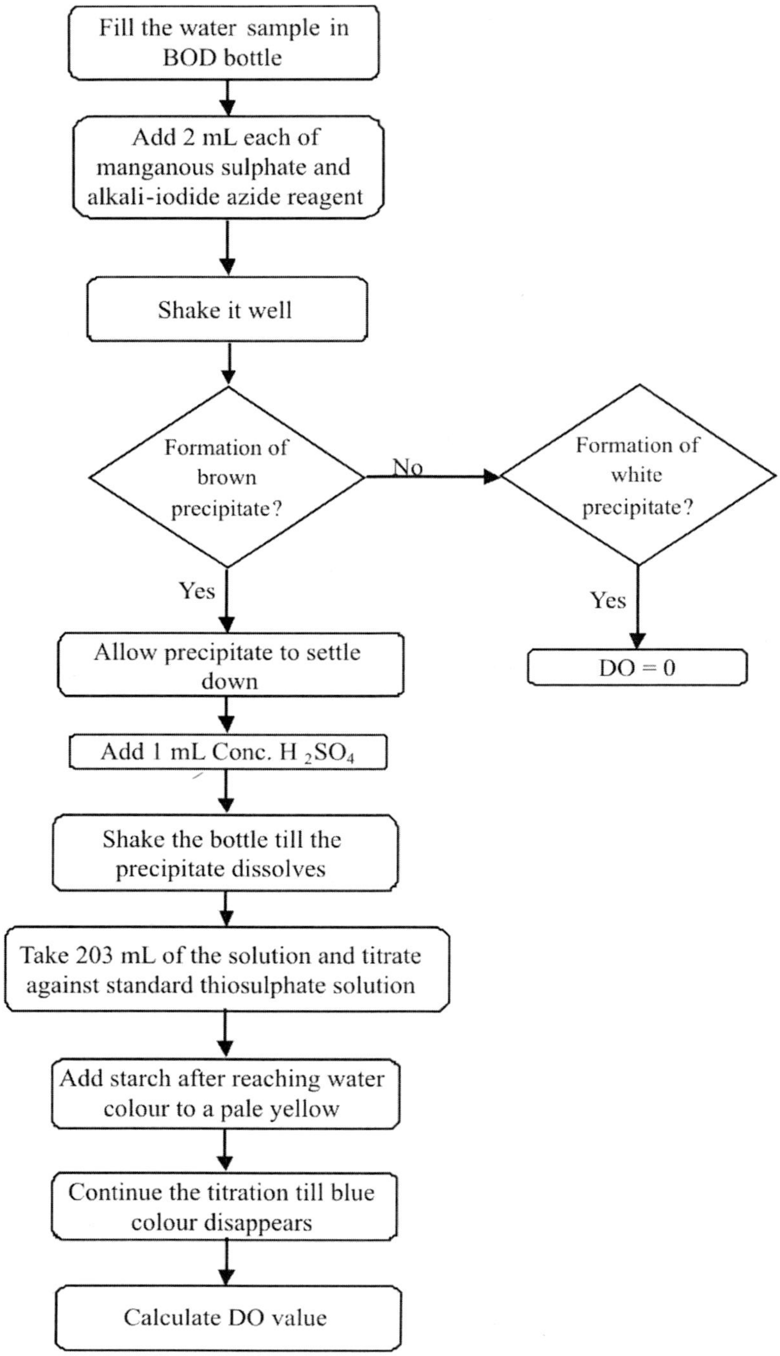

Fig. 7.1 Flowchart of DO determination procedure

TABULATION

Burette solution: Sodium thiosulphate solution (0.025 N)

Indicator: Starch

End Point: Blue to colourless

No.	Volume of Sample (mL)	Burette Reading (mL)		Volume of Sodium Thiosulphate (mL)
		Initial	Final	

QUESTIONS

1. Explain the factors which affect the DO concentration.
2. Find the theoretical DO saturation value at a water temperature of 28°C.
3. Explain self purification of river in detail.
4. What are the environmental significances of DO analysis?
5. Explain oxygen fixation reaction and its significance.
6. Vigorous shaking of samples is required after the addition of alkali iodide reagent and $MnSO_4$. Why?
7. Effective mixing of samples is required after the addition of sulphuric acid. Why?
8. What are the interferents in Winkler method? Explain their reactions.
9. In Winkler method, the blue colour of the starch indicator return even after the addition of higher amount of sodium thiosulphate. What is the reason behind this? How can we rectify this?
10. What is the reason behind the lesser dissolved oxygen concentration in water bodies during summer?
11. River water samples were taken from the same spot and filled in two BOD bottles for acquiring DO concentration. To the one BOD

bottle, 2 mL each of manganous sulphate and alkali-iodide-azide reagents were added at the same time of sample collection. Both samples are analyzed in the laboratory. What would be the results from both the samples? Are they same? If not, what are the possible reasons behind this?

12. What is the need of diluting the sample (colour changes from brown or dark yellow to pale yellow) before adding starch indicator?

OBJECTIVE QUESTIONS

1. Minimum DO level for the survival of aquatic life is.......................
2. Natural diffusion of air in surface water follows .... law.
3. ....in water may shoot-up the DO concentration in a high level.
4. The maximum possible DO concentration in water is .... mg/L.
5. Presence of carp, mosquito and black-fly larva etc. in water indicates the ....zone of river.
6. The zone of river at which DO concentration is very less, or zero is known as ....
7. Presence of worms, larva of mosquito and midge etc. indicates .... zone
8. The zone of river showing the increasing DO level is known as ....zone.
9. The zone of river with minimum pollutant concentration is known as .... zone
10. ....method is widely used to establish the DO concentration present in the water sample.
11. The presence of white precipitate with the addition of $MnSO_4$ indicates ....
12. The addition of divalent manganese ions in water is known as .... reaction.
13. .... medium is required for the liberation of iodine gas.
14. Starch indicator can be preserved by adding .... during its preparation
15. .... suppress nitrate interference.
16. After the fixation reaction, the sample can store up to .... hrs.

❑❑❑

8

Available Chlorine in Bleaching Powder and Test for Residual Chlorine

INTRODUCTION

Disinfection is the process of deactivating or destroying pathogenic microorganisms present in water. The agent responsible for the disinfection process is known as disinfectants. Disinfectants kill the pathogens either by damaging the cell wall or altering colloidal nature of protoplasm, cell permeability and organisms DNA or RNA. Various physical and chemical disinfectants such as ozone, chlorine, bromine, UV rays, lime, copper, silver, ultrasound, heat, potassium permanganate, hydrogen peroxide etc. has been proved as very efficient. Among these disinfectants, chlorine is used worldwide for the disinfection process. This is mainly due to the higher residual protection capacity of chlorine compared to other disinfectants.

Molecular chlorine, calcium hypochlorite and sodium hypochlorite are the main sources of chlorination. These compounds produce hypochlorous acid (HOCl) or hypochlorite (OCl^-) in water according to the pH of water. All the chlorine is in the form of hypochlorous acid at pH below 5 and in the form of hypochlorite ion at pH greater than 10. Thus disinfection activity of chlorine is by the hypoclorous acid, hypochlorite and chlorine gas present in water.

Bleaching powder [Calcium hypochlorite $Ca(OCl)_2$] is a commonly used disinfectant. The addition of bleaching powder produces HOCl or OCl^- in water according to water pH. Reducing agents such as ferrous ion, sulphide, manganese and nitrite converts chlorine to chloride. This reduces the disinfection activity of chlorine added in water. Further it reacts with organic compounds present in water in various ways. Thus, the added chlorine has no disinfecting capacity up to the removal of organic compounds and conversion of chlorine to chloride by reducing agents. Further addition of chlorine in water reacts with ammonia and forms chloramines- monochloramine, dichloramine and trichloramine according to the concentration of chlorine. Monochloramine and dichloramine has some disinfecting capacity compared to trichloramine. These compounds are longer lasting and give some degree of residual protection. Further addition of chlorine in water containing monochloramine and dichloramine converts these compounds to trichloramine, nitrogen gas and nitrous oxide. At these conditions, the added chlorine has no residual concentration. This trend continues up to the weight ratio of ammonia to chlorine reaches to around 7.6. After this, the added chlorine converts as residual chlorine and the point corresponding this is known as break point. At the break point, all the reducing agents, organic compounds and ammonia are converted to other forms and these converted compounds don't create any harm to the disinfecting activity of chlorine.

The addition of chlorine beyond the breakpoint is known as breakpoint chlorination. The residual chlorine concentration and added chlorine concentration has a direct relationship after the breakpoint. A residual protection of water is required even after the proper disinfection process. This is mainly because of reducing the attack of pathogens during the water supply. A residual concentration of 0.2 to 1 mg/L required for this purpose and a minimum of 0.5 mg/L is required for the protection from viral infection [12]. Thus, the amount of chlorine added to reach this residual protection is known as chlorine demand.

AIM

The aims of the experiment are:

a. To determine the amount of available chlorine in the bleaching powder by iodometric method.

b. To find the chlorine dosage for the given sample of water using orthotolodine method.

c. To find the residual chlorine present in water by iodometric method.

PRINCIPLE

Iodometric Method: At acidic conditions, chlorine reacts with potassium iodide and generates iodine gas (Eq. 8.1). The liberated iodine can quantify by starch-iodide titration method. The liberated iodine is titrated with a standard sodium thiosulphate ($Na_2S_2O_3$) solution in the presence of starch indicator (Eq. 8.2). The disappearance of blue colour indicates the completion of reaction with free iodine converted to iodide.

$$Cl_2 + 2KI \rightarrow 2KCl + I_2 \tag{8.1}$$

$$I_2 + 2Na_2S_2O_3 \rightarrow 2\ NaI + Na_2S_4O_6 \tag{8.2}$$

Orthotolodine Method: It measures both free and combined available chlorine. To obtain the correct colour development with chlorine and orthotolodine, (a) the solution must be at pH 1.3 or lower during the contact period (b) the ratio by weight of orthotolodine to chlorine must be at least 3:1 and, (c) the chlorine concentration must not exceed 10 mg/L. The orthotolodine reacts with residual chlorine and give yellow colour to the sample. The intensity of colour formation is proportional to the amount of residual chlorine present. Usually permanent colour standards representing different values of chlorine residuals will be prepared and kept ready for comparison.

REAGENTS PREPARATION

Standard iodine solution of normality 0.025N can be prepared by dissolving 3.25 g of resublimed iodine after dissolving 10 g of potassium iodide in chlorine-demand-free water of volume 25 mL and dilute to 1L. Orthotolodine solution can be prepared by dissolving 1.35 g of orthotolodine dihydrochloride in 500 mL distilled water and mixing this solution to a mixture of 350 mL distilled water and 150 mL concentrated HCl under constant stirring. Store the above mentioned solutions (iodine and orthotolodine) in brown bottles. Preparation of starch indicator and standard sodium thiosulphate has been discussed earlier.

PROCEDURE

A. Available Chlorine in Bleaching Powder using Iodometric Method

1. Take 1 g of bleaching powder and make a paste by adding a few amount of distilled water. Dissolve this paste in 1 L of distilled water.
2. Take 'V' mL of this solution and add 5 mL acetic acid and 1 g potassium iodide.

The addition of acetic acid reduces the pH of water to around 3 to 4. This creates a suitable condition for the iodine liberation reaction. Even though iodine liberates at pH below 8, pH 3 to 4 is the very suitable condition. This is mainly because of the partial oxidation of thiosulphate to sulphate at the neutral conditions.

3. Titrate the above solution with 0.025 N sodium thiosulphate until the solution colour turns to pale yellow. Add starch solution and continue the titration till the disappearance of blue colour. Note down the volume of sodium thiosulphate solution added (A).

4. Take a volume of distilled water corresponding to the volume of sample used. To this distilled water add 5 mL acetic acid, 1g potassium iodide and 1 mL starch solution.

   Blank correction is required to care of the oxidising or reducing reagents impurities and the iodine bound to starch at the end point.

5. If blue colour appears after the addition of starch, titrate with 0.025 N sodium thiosulphate solution until the blue colour disappears and note down the volume of sodium thiosulphate solution added (B_1).

6. If no blue colour occurs after the addition of starch, titrate with 0.025 N iodine solution until a blue colour appears. Note down the volume of iodine (B_2).

7. Then, titrate with 0.025 N sodium thiosulphate solution till the blue colour disappears. Record the volume of sodium thiosulphate solution added (B_3). Note down the difference between the volume of iodine solution and sodium thiosulphate as B_4 ($B_4 = B_2 - B_3$).

8. Calculate the percentage of chlorine present in water as follows

$$\text{Concentration of chlorine (mg/L)} = \frac{(A-B_1)\ \text{or}\ (A+B_4) \times 0.25 \times 35.45}{\text{Volume of bleaching powder solution (mL)}} \quad (8.3)$$

1000 mL of bleaching powder contains = 1000 x .... mg of chlorine

i.e. 1000 mg of bleaching powder contains .... mg of chlorine

Percentage of chlorine available in the bleaching powder (X) = ...

B. Determination of Chlorine Dosage for the Given Water Sample using the above Bleaching Powder Solution

1. Take 200 mL of the given water sample into 5 flasks. Add suitable amount of bleaching powder solution to the flasks. The concentration of chlorine added may be as 0.1 mg/L, 0.2 mg/L, 0.5 mg/L, 1 mg/L and 1.5 mg/L. Mix the solution thoroughly.
2. After 30 min of contact period, pour the chlorine added water to the middle chamber of the cell of the chloroscope. Add 5 drops of orthotolodine solution to this sample and mix well with the plunger. Fill the outer chambers with unchlorinated water sample and leave for 5 min.
3. Put the glass disc which indicates the dose of chlorine on its top, in front of the outer chamber. Compare the colour of the middle chamber with the colour of the glass disc. The disc which matches with the colour of the sample will give the residual chlorine present in the water.
4. Select the lowest dose which gives 0.2 mg/L of residual chlorine as chlorine dosage (Y).
5. Find the chlorine demand and amount bleaching required as follows.

Chlorine demand = Amount of chlorine added – 0.2 mg/L residual chlorine

$$\text{Amount of bleaching powder required per litre of water sample} = \frac{100Y}{X} \tag{8.4}$$

C. Determination of Residual Chlorine Present in the Water Sample using Iodometric Method

1. Follow the procedure step 2 and 3 of section A (Available Chlorine in Bleaching Powder using Iodometric Method)
2. The residual chlorine present in water is calculated as

$$\text{Residual chlorine present in water (mg/L)} = \frac{A \times 0.25 \times 35.45}{\text{Volume of water sample (mL)}} \tag{8.5}$$

TABULATION

A. Iodometric Method

Bleaching powder solution × Standard sodium thiosulphate solution (0.025 N)

Burette solution: Sodium thiosulphate

Indicator: Starch

End Point: Blue to colourless

No.	Volume of Bleaching Powder (mL)	Burette Reading (mL)		Volume of Sodium Thiosulphate 'A' (mL)
		Initial	Final	

Distilled water × Standard sodium thiosulphate solution (0.025 N)

Burette solution: Sodium thiosulphate

Indicator: Starch

End Point: Blue to colourless

No.	Volume of Distilled Water (mL)	Burette Reading (mL)		Volume of Sodium Thiosulphate 'B_1'(mL)
		Initial	Final	

Distilled water × Standard iodine solution (0.025N)

Volume of iodine added (B_2) = ... mL

Burette solution: Sodium thiosulphate

Indicator: Starch

End Point: Blue to colourless

No.	Volume of Distilled Water (mL)	Burette Reading (mL)		Volume of Sodium Thiosulphate 'B_3'(mL)
		Initial	Final	

B. Orthotolodine Method

No.	Chlorine Added (mg/L)	Bleaching Powder Added (mL)	Residual Chlorine (mg/L)

C. Iodometric Method

Burette solution: Sodium thiosulphate

Indicator: Starch

End Point: Blue to colourless

No.	Volume of Water Sample (mL)	Burette Reading (mL)		Volume of Sodium Thiosulphate 'A' (mL)
		Initial	Final	

QUESTIONS

1. What is the difference between disinfection and sterilisation?
2. Explain electro-katadyn process in detail.
3. Explain various chlorination methods in detail.
4. Explain the effect of pH on chlorination.
5. Why do we prefer chlorination over other methods of disinfection?
6. What is the use of acetic acid addition in iodometric method?
7. Explain Chick's law.

OBJECTIVE QUESTIONS

1. Disinfection is the process of deactivating or destroying ....present in water.
2. All the chlorine is in the form of ....at pH below 5.
3. Chemical name of bleaching powder is ....
4. At the breakpoint, ammonia to chlorine ratio is ....
5. Minimum residual concentration from the viral infection is ....
6. The amount of chlorine added to reach the required residual protection is known as.......................

❑❑❑

9

Chemical Oxygen Demand

INTRODUCTION

Chemical Oxygen Demand (COD) test is a widely used test for finding the organic strength of water and wastewater. COD can be defined as the amount of oxygen required for the complete oxidization of organic matter present in water. All the organic compounds present in water are converted to carbon dioxide, water and ammonia with strong oxidizing agents under acidic conditions as in the Eq. 9.1 [20].

$$C_nH_aO_bN_c + \left(n+\frac{a}{4}-\frac{b}{2}-\frac{3c}{4}\right) O_2 \rightarrow n\,CO_2 + \left(\frac{a}{2}-\frac{3c}{2}\right) H_2O + cNH_3 \quad (9.1)$$

The ratio of biochemical oxygen demand (BOD) to COD is very useful for the selection of wastewater treatment units. BOD is the measure of oxygen requirement for the degradation of organic matter by microorganisms. The high BOD values indicate the higher biodegradability of organic matter present in water. Thus, the BOD/COD ratio greater than 0.5 can be treated easily by biological methods and the ratio less than 0.5 should be treated by physico-chemical methods.

The value of COD is always higher than that of BOD. Because chemical oxidizing agents used in COD analysis can oxidize the biologically resistant organic compounds completely. For example, lignin is completely oxidized during the COD analysis. Thus, the BOD/COD ratio decreases with increase in biologically resistant organic matter concentration. The value of COD is higher than 2.5 times of BOD for domestic and some industrial wastewater.

The COD test can be completed within 3 hr. This is one of the major advantages of COD over BOD analysis. But, it is very difficult to differentiate between biologically oxidizable and biologically inert organic matter by COD test [14].

PRINCIPLE

Due to the higher ability to oxidize organic matter and relatively cheap, dichromate is widely used as the oxidizing agent in COD analysis. Dichromate can oxidize a wide variety of samples in the presence of H_2SO_4, as given in Eq. 9.2. The oxygen transaction from dichromate ion to carbon dioxide, results in the reduction of hexavalent chromium ion to trivalent chromium ion.

$$C_nH_aO_bN_c + d\,Cr_2O_7^{2-} + (8d + c)H^+ \rightarrow n\,CO_2 + \left[\frac{a+8d-3c}{2}\right] H_2O + cNH_4^+ + 2d\,Cr^{3+} \quad (9.2)$$

$$\text{Here } d = \frac{2n}{3} + \frac{a}{6} - \frac{b}{3} - \frac{c}{2}$$

The excess dichromate ions remaining after the reaction is titrated with ferrous ammonium sulphate (FAS), using ferroin as indicator (Eq. 9.3). The end point of this reaction is indicated by a colour change from blue-green to a reddish-brown. The dichromate consumed by the sample is equivalent to the amount of oxygen required to oxidize the organic matter.

$$6Fe^{2+} + Cr_2O_7^{2-} + 14H^+ \rightarrow 6Fe^{3+} + 2Cr^{3+} + 7H_2O \quad (9.3)$$

REAGENTS PREPARATION

Potassium dichromate solution of 0.01667 M is the source of oxidant for the COD analysis. Take 4.903 g $K_2Cr_2O_7$ in a 1L standard flask. Potassium chromate should be previously dried at 150°C for 2 h. Add 107 mL concentrated sulphuric acid to this salt and mix it well. Take 60 mL of concentrated H_2SO_4 in a beaker. Add 33.3 g $HgSO_4$ into the beaker and mix it well. Mix this solution into the standard flask. Keep the standard flask in a water bath during these operations. Because, the above process is highly exothermic in nature. Add distilled water to this solution and make it up to 500 mL. Keep the flask in the water bath and after one day make up to 1 L by adding distilled water.

Ferroin indicator solution is prepared by dissolving 1.485 g of 1,10-phenanthroline monohydrate and 695 mg of $FeSO_4.7H_2O$ in distilled water. Dilute this solution to 1L. Sulphuric acid reagent is prepared by dissolving silver sulphate at a rate of 5.5 g Ag_2SO_4/kg H_2SO_4. Silver sulphate is not easily dissolvable in sulphuric acid. Therefore, keep the solution for one to two days until dissolving the silver sulphate salt.

Standard ferrous ammonium sulphate titrant (0.1 M) can be prepared by dissolving 39.2 g $Fe(NH_4)_2(SO_4)_2.6H_2O$ in distilled water. To this solution, add 20 mL concentrated H_2SO_4. Cool the solution to room temperature and dilute to 1L.

PROCEDURE

Closed reflux and open reflux methods are the two methods available for the COD determination. Between these two methods, open reflux method is appropriate for variety of wastes with substantial sample size. In closed reflux method, pre-measured reagents are taken in COD vials and kept for digestion. This method is more economical than open reflux method, since the quantity of reagents are too small. Complete oxidation of volatile organic compounds take place in closed reflux method than open reflux method. This is mainly due to the longer contact time with oxidants in closed reflux method.

1. Take 2.5 mL sample in a COD vial; and add 1.5 mL $K_2Cr_2O_7$ reagent and 3.5 mL H_2SO_4 reagent carefully. Similarly take 2.5 mL of distilled water in another COD vial and add the above mentioned regents.

2. Keep these vials in COD digester and switch on the instrument. Keep the vials in the digester for 2h at 150°C.

3. After the digestion period, cool the samples to room temperature. After the digestion period, the sample should be in yellow colour. Otherwise, the sample is not suitable for titration and should be discarded.

4. Pour the mixture in conical flask and add ferroin indicator to this solution. The colour of the solution turns to green with the addition of ferroin indicator.

5. Titrate it with standard FAS solution until the end point reaches. The end point of this titration is the sharp colour change from blue-green to reddish brown. Sometimes, the initial colour may reappear within minutes.

6. Repeat the same procedure for blank solution and calculate the COD of given sample from the following equation.

$$\text{COD in mg/L} = \frac{(A\text{-}B) \times N \times 8 \times 1000}{\text{Volume of sample taken (mL)}} \quad (9.4)$$

where, A = Volume of FAS used for blank (mL); B = Volume of FAS used for sample (mL); N = Normality of FAS.

Here, blank correction is used to find the un-reacted dichromate concentration in the absence of water sample. From the above equation (Eq. 9.3), 6 mole of FAS is required for the complete reaction with one mole of potassium dichromate. In the standard analysis, 1/6 mole of potassium dichromate is used to find the COD value. The corresponding FAS required for the complete reaction is 1M. and 0.1 M FAS is used in the standard analysis. Thus, 10 mL of 0.1M FAS solution is required for the complete reaction of 1 mL of 1/6 M potassium dichromate solution. The volume of dichromate solution in the COD analysis is 1.5 mL. Thus, the volume of FAS required for the complete reaction is 15 mL.

Chloride is the main interferent for the COD analysis. The chloride ions convert dichromate to trivalent chromium in acidic medium as in Eq. 9.5. Mercuric sulphate addition is the best way for the elimination of this chloride interference. Mercuric ions react with chloride ions and form mercuric chloride complex as in Eq. 9.6. [14]. This mercuric sulphate is added in standard potassium chromate solution during its preparation.

$$6Cl^- + Cr_2O_7^{2-} + 14H^+ \rightarrow 3Cl_2 + 2Cr^{3+} + 7H_2O \quad (9.5)$$

$$Hg^{2+} + 2Cl^- \Leftrightarrow HgCl_2 \quad (9.6)$$

Dichromate ions are very weak to oxidize some of the organic compounds even at high acidic conditions. Fatty acids having low molecular weight are one of the examples of such organic compound. Dichromate can oxidize these organic compounds in the presence of a suitable catalyst. Silver ion is found as the suitable catalyst for the degradation of these organic compounds [14]. Thus, silver ions are incorporated in sulphuric acid reagent.

The COD analysis procedure is also depicted in Fig. 9.1.

TABULATION

Burette solution: Ferrous ammonium sulphate

Indicator: Ferroin

End Point: blue-green to reddish brown

No.	Volume of Sample (mL)	Burette Reading (mL)		Volume of FAS (mL)
		Initial	Final	

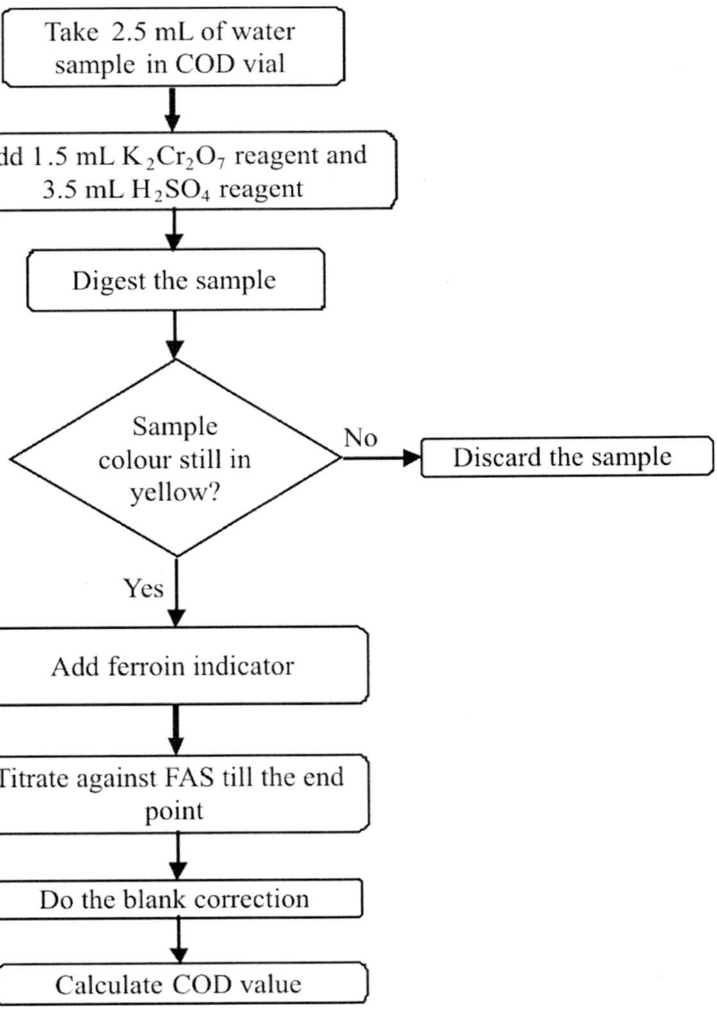

Fig. 9.1 Flowchart of COD analysis procedure

QUESTIONS

1. What are the advantages and disadvantages of COD analysis?
2. Which is the main interferents in COD analysis? How it can be prevented?
3. What is the use of silver sulphate addition during the preparation sulphuric acid reagent preparation?
4. Explain the principle of dichromate reflux method.
5. What is the theoretical FAS volume required for the blank correction? Prove it.

OBJECTIVE QUESTIONS

1. ....is the measure of oxygen requirement for the degradation of organic matter by microorganisms.
2. The value of ....is always higher than that of BOD.
3. BOD/COD ratio decreases with increasing ....concentration.
4. The value of COD is higher than .... times of BOD for the domestic wastewater.
5. .... is the indicator in COD analysis.
6. .... interfere COD analysis.
7. .... is a catalyst for the oxidation of organic matter.
8. The theoretical volume of FAS for the blank analysis is ....
9. .... reduces the interference of choride ion.
10. The .... convert dichromate to trivalent chromium in acidic medium

❑❑❑

10

Answers (Objective Questions)

CHLORIDE

1. 250 mg/L
2. 1000 mg/L
3. Argentometric
4. Pinkish yellow, brick red
5. 3.5%
6. Higher
7. Lakes (Shallow lakes)
8. Magnesium chloride
9. Tracer
10. Chloride
11. 30000 mg/L
12. Calcide
13. 6g per person per day
14. Hydrogen peroxide
15. 7 to 8
16. Silver chloride
17. Sodium chloride
18. Chloride free distilled water

ALKALINITY

1. Alkalinity
2. Bitter
3. Very corrosive
4. Bicarbonate
5. 0.02N
6. 10.5
7. 8.3
8. 4.5
9. Caustic alkalinity
10. Bromocresol green and methyl red
11. Phenolphthalein
12. Colourless
13. Blue
14. Red

ACIDITY

1. Acidity
2. Below 4.
3. 3.7
4. Orange
5. Mineral acidity
6. Bromphenol blue
7. Yellow to purple
8. Metacresol purple

HARDNESS

1. Scum
2. Chelating
3. Laxative
4. Murexide
5. Carbonate hardness
6. Carbonate/temporary hardness
7. Pseudo-hardness
8. Cardio vascular
9. Wine red
10. Ammonium chloride and ammonium hydroxide

DISSOLVED OXYGEN

1. 4 mg/L
2. Henry's law
3. Algal bloom
4. 14.652 mg/L
5. Degradation
6. Septic zone
7. Septic zone
8 Recovery
9. Clear
10. Winkler method
11. Absence of DO (or formation of $Mn(OH)_2$)
12. Oxygen fixation reaction
13. Acid
14. Salicylic acid
15. Azide
16. Eight

AVAILABLE CHLORINE IN BLEACHING POWDER AND TEST FOR RESIDUAL CHLORINE

1. Pathogenic microorganisms
2. Hypochlorous acid
3. Calcium hypochlorite $Ca(OCl)_2$
4. 7.6
5. 0.5 mg/L
6. Chlorine demand

CHEMICAL OXYGEN DEMAND

1. BOD
2. COD
3. Biologically resistant organic matter
4. 2.5
5. Ferroin
6. Chloride
7. Silver ions
8. 15 mL
9. Mercury ions
10. Chloride ions

❑❑❑

References

1. P. R. Sreemahadevan Pillai, (2009) A Comprehensive Laboratory Manual for Environmental Science and Technology, New Age International (P) Limited, Publishers, New Delhi.
2. Science lab.com; Material Safety Data Sheet, Nitric acid.
3. Science lab.com; Material Safety Data Sheet, Hydrochloric acid.
4. Science lab.com; Material Safety Data Sheet, Sulphuric acid.
5. Science lab.com; Material Safety Data Sheet, Sodium Hydroxide.
6. Science lab.com; Material Safety Data Sheet, Potassium chromate
7. Science lab.com; Material Safety Data Sheet, Silver nitrate.
8. Science lab.com; Material Safety Data Sheet, Acetic acid.
9. Aditya Birla Chemicals (India) Limited; Material Safety Data Sheet, Bleaching Powder.
10. Benfield LD, Weand BL, Judkins JF: Process chemistry for water and wastewater. New Jersey: Prentice Hall Inc, Englewood Cliffs; 1982.
11. Garg S.K. Environmental Engineering. Vol. I Khanna Publishers, New Delhi.
12. IS 10500: 2012 Indian Standard Drinking Water—Specification. Bureau of Indian Standards, New Delhi.
13. S.N. Davis, G. M. Thompson, H. W. Bentley, G. Stiles, Ground Water Traces: A Short Review; Groundwater 18 (1980) 14-23.
14. Sawyer, C.N., McCarty, P.L. and Parkin, G.F., *Chemistry for Environmental Engineering*, Tata McGraw-Hill, New Delhi, 2003.
15. Korbi's Laboratory - Precipitation and reduction of thiosulfate with silver nitrate; http://www.korbis-labor.de/experimente/faellung/en-thiosulfat. htm. Accessed on 28.10.2015.

16. Nuffield Foundation, Catalysis of the reaction between sodium thiosulphate and hydrogen peroxide. http://www.nuffieldfoundation. org/practical- chemistry/catalysis-reaction-between-sodium-thiosulfate-and- hydrogen-peroxide.

17. http://www.chemguide.co.uk/physical/acidbaseeqia/indicators.html

18. American Society of Civil Engineering Committee on Sanitary Engineering Research.1960. Solubility of atmospheric oxygen in water. J. Sanitary Eng. Div. 86(7): 41–53.

19. http://www.titrations.info/iodometric-titration-end-point-detection.

20. https://en.wikipedia.org/wiki/Chemical_oxygen_demand.

❑❑❑

About the Author

Dr. P. V. Nidheesh completed B. Tech. in Civil Engineering from Government College of Engineering, Kannur, Kerala; M. Tech. in Environmental Engineering and Ph. D. from National Institute of Technology Trichy, Tamilnadu. After the completion of doctoral degree, he worked as an adhoc faculty at National Institute of Technology Calicut, Kerala and as an Associate Professor at Vimal Jyothi Engineering College, Chemperi, Kannur, Kerala. Currently he is working as a senior scientist at CSIR-National Environmental Engineering Research Institute, Nagpur. His research interest includes, water and wastewater treatment, advanced oxidation processes and solid waste management. He has published 55 international journal articles, 9 conference articles and one book chapter. He is an author of highly cited and highly downloaded articles. He is a reviewer of various journals published by Nature, Wiley, Springer, Elsevier etc.